MAYOR'S DESK

MAYOR'S DESK

ANTHONY FLINT

LINCOLN INSTITUTE
OF LAND POLICY

Cambridge, Massachusetts

CONTENTS

CONTENTS

FOREWORD

MICHAEL R. BLOOMBERG

"WHEN THE BURDENS OF THE PRESIDENCY SEEM UNUSUALLY HEAVY," Lyndon Johnson said in 1966, "I always remind myself it could be worse. I could be a mayor." He was only half kidding.

Johnson understood the extraordinary day-to-day pressures that mayors face as the elected officials most directly responsible for the public's well-being. And what was true in 1966 is even more true today. When it comes to the big challenges facing our nation and our world, the buck increasingly stops on the mayor's desk.

When Johnson spoke those words, about one in three people around the world lived in cities. Today, more than half do. By mid-century, it may be nearly 70 percent. As city leaders find ways to provide more services to more people, they are also confronting unprecedented challenges posed by climate change.

The good news is: more and more mayors are turning local govern-ments into laboratories of innovation and invention, as these 20 conversations help show. A common thread throughout these pages is the role of land—namely, how mayors are rethinking the way it is used to tackle poverty, climate change, housing affordability, transportation, and much more. Each conversation offers a window into how urban ideas—from transformational projects to more nuts-and-bolts policies—take shape and seek to improve the lives of citizens.

This book is designed for those citizens, to help highlight the power and promise of local government—but also for local elected officials around the world who recognize the potential that lies in collaboration.

Cities of all sizes face many common challenges. Mayors can learn from and adapt ideas that have been tested—and proven to work— in other places. That's something our administration tried to do in New York City during our 12 years in City Hall.

On many pressing issues, including climate change, transportation, and public health, we borrowed and adapted innovative new ideas from other cities, from Bogotá to Amsterdam to Tokyo. Our foundation, Bloomberg Philanthropies, has long been helping more cities do the same. Through a variety of programs that focus on strengthening mayoral leadership, increasing the capacity of city halls to tackle big issues, and deploying data more effectively, we work with city leaders from Cleveland to Cape Town to Copenhagen and beyond. This book aligns with another part of our mission: highlighting promising ideas and best practices, and helping to spread them around the world.

For mayors, activists, urban planners, students, and citizens of every kind, these pages offer a sample of some of the bold ideas that have been emerging from cities over the past decade. The mayors on these pages have differing political viewpoints and party memberships, and that underscores one of the book's messages: just as good ideas transcend national borders, they transcend political ideology, too.

The conversations include candid reflections on the difficulties of municipal leadership. Not every idea will work. Not every program will succeed. And that is OK—because failure is a learning experience and can be the basis of future success.

We need more leaders who are willing to test new ideas. Because it's only when we reject the status quo and take risks that we can make a difference. And the more that voters recognize the potential of that innovative spirit, the more they will demand it.

As President Johnson understood, being a mayor is an incredibly difficult job. But it's also where public servants with great ideas— and some courage—can have an enormous impact on their communities and beyond.

Michael R. Bloomberg is the founder of Bloomberg LP and Bloomberg Philanthropies. He served as mayor of New York City from 2002 to 2013, and chairs C40, a global network of mayors taking action on climate change.

SYRACUSE

BEN WALSH

Describing the use of data to improve snow removal for a national What Works Cities video, 2022.

Ben Walsh was only a few weeks into his first term as the 54th mayor of Syracuse, New York, when he sat for this interview, describing his plans for leading a postindustrial metropolis that, like many, had struggled with job and population loss. Prior to becoming mayor—a job held in the 1960s by his grandfather, who later served in Congress, as did his father—Walsh was involved with several downtown redevelopment projects as the city's deputy commissioner of Neighborhood and Business Development and head of the Syracuse Industrial Development Agency. Walsh began a second term in 2022, renewing his focus on strengthening schools, reducing gun violence, creating a Housing Trust Fund, and updating zoning and permitting rules to accommodate anticipated population and economic growth.

ANTHONY FLINT: You were born into a political family—your grandfather was mayor and congressman, and your father was also elected to Congress—but you took some time before running yourself. What finally prompted you to want to be chief executive?

BEN WALSH: It's the public service that is important. I admired my dad's ability to see politics as a means to an end. I was never confident in my own ability to strike that balance. But ultimately, when I came to work for the city under the previous mayor, Stephanie Miner, I started to consider running for office.

AF: In your first weeks on the job, what would you describe as the biggest challenge, and the biggest promise, in leading a legacy city such as Syracuse?

During Walsh's two terms, he has successfully worked with city leaders to stabilize the city fund balance, which was projected to exceed $100 million by the end of June 2023.

BW: We have a structural deficit—we're anticipating a $20 million shortfall this year. The good news is we have built up a good fund balance over the years. The bad news is we are drawing down on it. But that's balanced by all the ingredients that Syracuse has to become a vibrant city. I look at the trends and where young people want to be. We can offer them urban amenities, proximity to work, density, and walkability.

AF: According to our recent report *Revitalizing America's Smaller Legacy Cities,* many postindustrial cities have a strong tradition of foundations, other nonprofits, and "eds and meds," or anchor institutions. What kind of partnerships are you developing?

BW: We actually don't have a large philanthropic base. We have local foundations, but they're on the smaller side. It's a blessing and curse of our industrial legacy. We've never relied on one industry or company. But yes, eds and meds are a significant part of the city. One of the nation's highest concentrations of colleges and universities is in this region and three great hospitals are major employers. St. Joseph's Hospital has very intentionally grown in a way that supports the neighborhood around it. Same for Syracuse University. We would like to be doing a better job of commercializing the technology that comes out of those institutions, which have a foundation of knowledge and expertise to create new companies and industries. . . . We have a lot of companies working in unmanned aerial systems (UAS) innovation, based on radar technologies going back to General Electric. With Carrier, manufacturing is gone, but their R&D is still here, and some great work is being done on indoor air quality technologies.

In what will be the largest private investment in New York State history, Micron Technology plans to build a $100 billion, 1,400-acre semiconductor manufacturing plant in Clay, a suburb just north of Syracuse. The project is expected to create nearly 50,000 jobs in the area.

Below: Syracuse had a population of more than 220,000 in 1950. Today about 145,000 people call it home.

AF: Syracuse made a bid for Amazon HQ2. Did you learn anything from that process about Syracuse's assets or shortcomings?

BW: Making the bid forced the region to work together and collaborate, and helped prepare us for future, perhaps more realistic, opportunities. We've seen our fair share of hard times over the years, and I think that has made us a risk-averse community. I liked the way we thought outside the box.

Above: The Erie Canal cut through Syracuse for more than a century after its construction in 1820, transforming a village of 250 people into an industrial powerhouse.

Facing page: When the canal was rerouted in 1924, the city paved Clinton Square and put up a parking lot. In 2001, the square was redesigned as a public park, a revitalization strategy that continues to inspire nearby residential and commercial redevelopment.

AF: History and a sense of place are important qualities for many regenerating legacy cities to cultivate. What are the urban "bones" of Syracuse that give it a competitive advantage in this regard?

BW: The renaissance of our urban core is being driven by the adaptive reuse of our industrial and historic building stock. People are looking for that sense of place, that authenticity. It's real. We're not trying to recreate a Main Street. We've used both federal and New York State historic tax credits. That has been a major driver of our redevelopment. When you look more broadly at the importance of history here, a few things come to mind— the Syracuse region was the birthplace of the Iroquois Confederacy, the birthplace of the women's suffrage movement, the hub of the salt industry, and an important stop on the Underground Railroad. We're embracing that history.

AF: Can you tell us about your interest and experience in land banking, and how that might translate to more equitable development?

Launched in 2012, the Greater Syracuse Land Bank returns vacant, underutilized, and tax-delinquent properties to productive use by acquiring and stabilizing them, and then selling them to responsible buyers for redevelopment.

BW: We saw land banking, at first, as an opportunity to help the city be more effective in collecting taxes. We made a policy decision in the past not to foreclose on delinquent properties, and that created an environment where there wasn't any accountability. Properties were falling vacant. We also wanted to be more intentional in how we dealt with these vacant lots, not as liabilities but as assets. What we had been doing was selling tax liens. We realized we owned the problem regardless. So we passed state legislation to create a city-county land bank. Since then, we've built up a sizable inventory of properties and have sold over 500

I look at the trends and where young people want to be. We can offer them urban amenities, proximity to work, density, and walkability.

to date. Now we turn to more equitable development. We favor selling to homeowners, and we're enabling affordable housing development using low-income tax credits. The first step was getting our arms around the problem, and now we're looking at planning processes.

AF: The era of urban renewal took a particular toll on cities like Syracuse. How important is the proposed dismantling of the Interstate 81 viaduct through downtown, and what can be done to make that project become reality soon?

BW: I've been a vocal proponent of removing the elevated portion of I-81 in favor of the community grid option. We have existing infrastructure to reroute through traffic around the city and accommodate traffic coming into the city through an enhanced street grid. There are primarily suburban interests that understandably see any alteration of the existing conditions as a threat. At the state level, it's taking longer than anyone expected. We're waiting on a draft Environmental Impact Statement (EIS). The options include replacing the viaduct, which would require it

Below: In 2023, state and local officials broke ground on a $2.25 billion project to dismantle the elevated Interstate 81 viaduct through downtown Syracuse, pursuing the community grid concept illustrated here to reconnect neighborhoods and improve mobility.

to be higher and wider, and a tunnel option. We're talking a difference in billions. The community grid option comes in at $1.3
billion; the least expensive tunnel option is $3.2 to $4.5 billion,
and goes up from there. That's a pretty big difference for a mile
and a half stretch. Even if we could afford to build the tunnel,
we don't need it. Eighty percent of the traffic already comes into
the existing grid. It's just bottlenecked at a couple of off-ramps.
This is a once-in-a-generation opportunity to right a past wrong.

AF: Not every city can be a tech hub
alternative to Silicon Valley. What is
it about Syracuse that could create
a niche?

BW: We can't compete in every
arena, so we identify areas where we
have core competencies, like indoor
air quality and unmanned aerial
systems. I believe we're the only city
in the country where you'll be able to
test drones beyond line of sight.

Below: A drone delivers COVID
tests to SUNY Upstate Medical
University Hospital in Syracuse as
part of a feasibility test conducted
by NUAIR, a nonprofit developing
Central New York's unmanned
aerial economy. Since 2016, New
York State has committed more
than $112 million to help develop
a 50-mile unmanned aerial test
corridor between Syracuse
and Rome.

AF: What will it take to get younger people to stay in Syracuse?
Do you have a target population size in mind?

BW: I don't have a target in mind. For decades, we lost population. The first step is stabilizing. We've done that, and we're right
around 140,000. When we peaked as a population, we didn't have
the suburban sprawl we have today. The overall population of the
region remained stable, but we were losing downtown residents
as people moved to the suburbs. But now, when you look at the
national trends, the city is what young people are looking for.

Syracuse native Ben Walsh graduated from Ithaca College and holds a master's
degree in public administration from the Maxwell School at Syracuse University.
Early in his career, he held positions at the Metropolitan Development Association
of Syracuse and Central New York, and at the law firm of Mackenzie Hughes LLP.
His accomplishments before becoming mayor included helping to establish the
Greater Syracuse Land Bank, serving as executive director of the Syracuse Industrial Development Agency, and working on redevelopment projects including
the Hotel Syracuse.

MANUEL VELARDE

Overlooking the busy Vía Evitamiento bus stop in Lima, 2020.

SAN ISIDRO, LIMA

Lawyer and politician Manuel Velarde was sworn in as the 20th mayor of San Isidro, a district within Lima, Peru, in January 2015. As mayor of San Isidro, Velarde commuted to City Hall by bicycle and focused on public safety, adding more public space to San Isidro's collection of parks, and "prioritizing pedestrians over cars," as he told a reporter in 2016—all objectives that were popular with younger residents and urban design aficionados, but sometimes met resistance from older, wealthier constituents. In the fall of 2018, at the end of his term, Velarde ran for mayor of metropolitan Lima, but lost to Jorge Muñoz—the mayor of the neighboring Miraflores district. Since leaving office, he has publicly identified himself as a "citizen and activist . . . determined to transform the country based on a green agenda."

Below: As the financial center of Lima, San Isidro sees an influx of workers and shoppers during the week.

ANTHONY FLINT: Governance structure affects the administration of large metropolitan regions and the quality of life for its citizens. Can you tell us about the challenges and opportunities of being part of the governance system in Lima?

MANUEL VELARDE: San Isidro is one of 43 districts run by the Metropolitan Municipality of Lima. Each district has its peculiarities. We are a 10-square-kilometer (3.86-square-mile) territory with approximately 60,000 residents. We are also the financial center of Peru. From Monday through Friday, around a million people come into San Isidro to work, shop, or do some other kind of task. It's a big challenge to accommodate this influx. The policies we apply are seen as cutting edge. We're in a position to offer better services, generating a better quality of life, but we face challenges—for example, the need for more public transportation. We must also coordinate constantly with other districts.

With more than 10 million residents, Lima is Peru's political and economic capital, and home to nearly a third of the country's populace.

AF: What are the major financial and planning challenges in San Isidro, and how is the municipality dealing with those challenges?

MV: Today the district is financed by two taxes: the property tax and the service tax. Both taxes, but principally the service tax, provide the revenue for all services. In certain parts of our country, noncompliance is a big problem. That's because residents don't feel they get what they pay for, due to poor management and corruption. There's a lack of trust in the local government. In San Isidro, however, around 90 percent of residents and businesses pay their taxes on time, and that allows us to generate public investment. Our budget is always limited, and we need to prioritize. For that, we develop planning strategies to maximize the impact of investments.

Peruvian politics has long been punctuated by periods of corruption and instability, and no president has completed a term in office since 2016.

AF: San Isidro is considered the financial center of Lima, if not Peru, and its population has a relatively high level of income for the region. To what extent does the municipality rely on land-based resources and financial tools such as the property tax or land value capture?

Land value capture is a financing approach in which private developers contribute to municipal expenses for infrastructure, affordable housing, and other public goods.

MV: At this time, land value capture is not within our competencies. We're attracting private investment and creating public-private partnerships and making sure those projects are aligned with our sustainable development policies. The problem in San Isidro is

that property is expensive, and we lack the population—particularly younger residents—to support that. We need affordable housing. One solution is that we have reduced the minimum size of an apartment from 200 square meters to 45, 60, and 80 square meters (from approximately 2,100 square feet to 500, 650, and 850 square feet) to attract new residents, especially young people. We've started construction of new housing investments in the financial center. This will allow people to walk to their jobs and will reduce traffic congestion. We are focused on transit-oriented development.

Above: Police officers in San Isidro enforce traffic rules in 2017 on the first day of a campaign to encourage respect for pedestrians.

Facing page: San Isidro installed a protected bike lane and traffic-calming measures along Calle Los Libertadores, one of several such projects spearheaded by Velarde.

AF: Your efforts to prioritize pedestrians and bicycles over cars have prompted fierce criticism, including an attempted recall. Do you feel you have successfully changed the culture in the public realm?

MV: When I was elected mayor, I promised the voters I would modernize San Isidro but keep it on a human scale. Our area has suffered dramatically from the intensive use of cars. Our district needed to be retrofitted for pedestrians and cyclists. We began with the idea that it's more affordable to live without a car, and that cars are having negative effects on the city and quality of life. Transforming underutilized land and areas dominated by cars, we have created public spaces that people would not have thought possible a short time ago. Of course, it meets resistance—any city undergoing these kinds of reforms will face resistance. But as citizens start to recognize that they can live in a better environment than before, that will change.

In the beginning, we created bike lanes and parking for bikes, and then we wanted to provide a public system of bikes. We wanted to promote intermodality and alternatives for short trips that are currently made by car, but which should be made by bikes or on foot. Our new bike-share system will stretch that policy. We've already signed the contract, and the implementation will be done soon. The operator is the same investor that recently revamped the bike system in Paris.

In 2018, CityBike Lima signed an agreement with Velarde's administration to install a bike-share system with 500 bikes and 50 docking stations across San Isidro, but the company encountered resistance from Velarde's successor.

AF: Expansion of the Lima Metro mass transit system is underway. How important is public transportation in San Isidro, and how does it fit in with your planning?

Any city undergoing these kinds of reforms will face resistance. But as citizens start to recognize that they can live in a better environment than before, that will change.

MV: An additional metro line is under construction right now. We have one line in operation, but it does not cross the district. We will have to wait around 10 years more for the next lines that pass through San Isidro. The new lines will be underground and funded by the national government. Investment in public transport is crucial to facilitate accessibility for residents and visitors. At the same time, we need better management of parking spaces. We don't have parking meters, so we're inducing demand because people can park for free on the streets. We need to be able to build an efficient parking payment system.

AF: You have partnered with IBM and others to make the district a "smart city." Can you identify a few ways that technology has improved quality of life?

MV: We have to be careful with the use of technology. Look at history. At one point, we were told that using a car was affordable and efficient, but it's had a huge impact in cities. We have been victims of the presence of cars in our environment and of thinking that the car was an absolute solution. We now know it is not, so we have to avoid becoming victims of any other kind of trap. Technology is useful, but we can't commit the same mistake. What we need, more than a smart city, is smart citizens who know how to live in the city of the future.

A couple of years ago, we worked on a contest sponsored by IBM, and they gave us advice. We want to help people with intermodality—to give people the tools to make their trips more efficient. That means providing up-to-date departure times and showing how you can connect to other modes—such as where the bike share is, and how far it is to walk. That is how I view the role of technology.

Facing page: Parque Bicentenario opened in San Isidro in 2020 to mark the 200th anniversary of Peru. Intended to be the "lungs of the city," the five-acre coastal park includes play areas, an organic garden, a bike path, and other features.

The new metro line Velarde refers to is Line 2, a multibillion-dollar, 17-mile east-west expansion of the network that is expected to open in 2024.

San Isidro was one of 16 cities selected to participate in IBM's Smarter Cities Challenge in 2015. The tech giant consulted on sustainable mobility and traffic congestion.

Born in San Isidro, Manuel Velarde graduated from the Pontifical Catholic University of Peru and earned master's degrees in law from the University of Pennsylvania and King's College London. Early in his career, he worked as a financial legal advisor in New York, Lima, and Brussels. The son of a Peruvian government minister, Velarde entered public service in 2003 as a legal counsel at the Ministry of Economy and Finance of Peru. He was appointed head of the National Superintendency of Tax Administration in 2009. From 2010 to 2015, Velarde taught at the School of Economics at the University of San Martín de Porres. He served as mayor of San Isidro from 2015 to 2018.

WARSAW

HANNA
GRONKIEW

ICZ-WALTZ

Honoring World War II resistance leader General Zbigniew Ścibor-Rylski on Długa Street, 2018.

A native of Warsaw, Mayor Hanna Gronkiewicz-Waltz made her mark on this city of 1.7 million people as the city's first female mayor, serving an unprecedented three terms from 2006 to 2018. During her tenure, Gronkiewicz-Waltz faced controversial issues including the restitution of properties seized under Nazi and Communist rule, but also won plaudits for her advocacy and focus on issues including climate change and LGBTQ+ rights. In 2019, shortly after leaving office, she was selected to chair the Climate-Neutral and Smart Cities mission of the European Commission's Horizon Europe research and innovation program. Now on the faculty of the University of Warsaw, she is frequently called upon by the media to provide commentary on social and political matters.

Right: A May Day celebration in Warsaw's Old Town district, which was reconstructed after World War II.

ANTHONY FLINT: Last year, the national government proposed expanding Warsaw by bringing more than 30 outlying districts within its boundaries, an idea you opposed. In your view, what are the merits of a more regional approach to metropolitan governance?

HANNA GRONKIEWICZ-WALTZ: The target of that proposal was purely political—one party saw the opportunity to get power in Warsaw through votes from around the region. They wanted to enlarge the municipality in an effort to get people from the countryside and the smaller towns to vote for the next mayor of Warsaw. We protested, and in various local referendums the people said no. They preferred remaining independent, with their own local governments and their own mayors.

People understand that our metropolitan policies have been successful. We collaborate as a region through contracts and agreements, and we rely on revenue sharing among the municipalities that make up our metropolitan area. Funding is organized through the Integrated Territorial Investments, an EU program, for these municipalities, with investment in everything from administration capacity to bike paths. That is the way to trust each other. And it works. There is an efficient public transportation scheme in place, under which the capital city's fleet serves the whole metropolitan area. Metropolitan governance should always respect the needs of all its members.

AF: What are the critical elements in your effort to maintain good municipal fiscal health? What has been your experience on the revenue side?

HG: On the revenue side, we have a property tax; it's not very high, though some people complain. We also have a lease tax, which is adjusted to the value of the property. A typical apartment tax bill in the city center is about $400 per year. There is also the commercial property tax and a tax on civil law transactions. However, these are only a small percent of the total budget. The biggest revenue source is the city's share in personal and corporate income tax, which flows directly from the central government. There are many needs for revenue; for example, we contribute to teachers' salaries, and we have to maintain our infrastructure.

Unlike most property taxes in Europe and the United States, property tax rates in Poland are based on area, not value. Warsaw's property tax rates in 2022 were $6.36 per square meter ($0.59 per square foot) for commercial properties and $0.22 per square meter ($0.02 per square foot) for residential properties.

Above: Electric buses have become more common as Warsaw replaces its diesel fleet.

The national Mój Prąd (My Electricity) program, introduced in 2019, offers subsidies to homeowners for solar panels. In 2022, in response to energy prices rising due to the war in Ukraine, the government expanded the program to include heat pumps, solar hot water collectors, and other energy-efficient devices. Since the program's launch, it has funded over 410,000 applications.

Between 2007 and 2020, Warsaw secured EU funds of nearly €2 billion ($2.1 billion) to finance the expansion of its metro system. The investment is part of a broader effort to develop low-emission public transportation across Europe.

AF: Speaking of infrastructure: how is climate change going to impact Warsaw, and what is the city doing with respect to mitigation and adaptation?

HG: The main fuel for so long was coal. Step by step, we have to move away from coal, and change to natural gas and renewables. First we focused on transportation—new buses, new trams, and a second metro line. We are changing our rolling stock, replacing diesel buses with electric and natural-gas models. Seventy percent of our citizens use public transport.

The modernization of our district heating network, which serves 80 percent of the city's residents, is also very important. Ten thousand additional homes have been connected to the system in the past 10 years. Warsaw's heat is produced in two combined heat and power plants. We are planning to switch one of the plants from coal to gas, which will bring a significant carbon dioxide emission reduction. Also, individuals can apply for subsidies to install photovoltaics, solar panels, and heat pumps, and thus replace old-fashioned stoves. This has been a very popular program, inspiring hundreds of applications. We are active internationally as well; for example, we are part of the EU Covenant of Mayors, which is committed to implementing climate and energy initiatives.

AF: What results have you seen from the expansion of public transit? Has it been successful in terms of ridership and reduced traffic congestion?

HG: As for being car-free, people know one day it will come, though it may have to come from my successors. The way it was done in London—starting with a pilot for one year—was very good. People there decided they preferred the congestion fee and supported the money going to transit. Public transport is costly. We have been able to accomplish so much because 85 percent of the investment was covered by EU funds.

For users, it's important for public transport to be cheap. In Warsaw, the price is about $30 per month, and our seniors pay $20 for the whole year. Last year, we began offering free transportation for students up to 15 years old; families need to teach the young that it's OK to go by bus. We have dedicated bus lanes, 500 kilometers

(310 miles) of bicycle lanes, and bike sharing. Even with all of that, there is still congestion, though it is not as bad as before.

AF: As Warsaw joins the array of economic powerhouses, how are you addressing gentrification, providing affordable housing, and fostering a more inclusive economy?

HG: We had to start from scratch. There was no private ownership under Soviet control. Beginning in the 1970s, there was a policy that let you buy your home for 10 percent of its value. I was the first mayor who stopped that kind of sale of municipal apartments. At the same time, we started to build more housing; we've added 3,500 new apartments over the last 10 years. We use the city's land and keep the construction costs down, so people's rents are not so high. I lived in Knightsbridge, in London, for a few years, and I saw how investments by foreign developers made the price of apartments skyrocket. We don't have that in Warsaw—housing prices are rising gradually, but at an affordable pace. Another problem is that many apartments have not been maintained. That is why the city is directing finances toward revitalization, especially in the most neglected neighborhoods.

Below: Dmowski Roundabout, a transit hub in Warsaw's city center.

Next page: Ukrainian refugees fill Warsaw's central train station in March 2022. Poland registered more than 1.5 million refugees in the first 10 months after Russia invaded Ukraine, with more than 11 million others crossing the Polish border as they fled to other destinations.

AF: What have been the effects of rising nationalism and anti-immigration sentiment on the city's economy, taxation, and social spending?

HG: The national government decided to withdraw from Poland's agreement to accept, under the EU's quota system, a proportional number of refugees. This was not helpful, as we have abandoned our European allies in the midst of the refugee crisis.

Generally, anti-immigration sentiment can discourage investment in the medium and long term. It's a very bad thing when someone with a different ethnic background is attacked on the bus, and it can also inhibit others from immigrating to Poland, including businesspeople. On the other hand, Warsaw does have many foreigners who come as economic migrants, and the majority of them are from Ukraine. Some are teachers, some are doctors; they are nannies or they work in the shops. We also have a significant number of Vietnamese immigrants, as well as people from Somalia, Ethiopia, and Chechnya.

To meet the growing need for integration, the city has created a multicultural center, which offers free language and cultural courses. It is important for the economy that we welcome and train immigrants; it helps new residents integrate into our society. As a consequence, the labor market is better off; the unemployment rate is 1.7 percent in Warsaw. The local economy is booming, as you can see from the city's many construction sites—they have to compete for workers. Economically, we are certainly benefiting from migration.

Facing page: New apartments under construction.

In 2015, the European Union created a plan to resettle 160,000 refugees from the Middle East and Africa among its member nations. Poland and Hungary refused to accept any of the migrants. In 2023, the two countries vetoed an EU statement of solidarity on migration.

In early 2023, Warsaw's unemployment rate was even lower, at 1.4%, according to the city's Statistical Office. Salaries were up over the prior year.

Hanna Gronkiewicz-Waltz earned a law degree and doctorate of legal sciences from the University of Warsaw, where she later became a professor of law and economics. She has served as president of the National Bank of Poland and vice president of the European Bank for Reconstruction and Development in London. The Warsaw native has also been a member of the Polish Parliament, chair of the State Treasury Commission, and president of Eurocities, a network of over 200 major European cities. Gronkiewicz-Waltz served as mayor of Warsaw from 2006 to 2018.

GLEAM DAVIS

City Hall, Main Street, 2018.

SANTA MONICA

For some people, Santa Monica conjures images of sunshine and surfing. But the Southern California city should rightly be known for sustainability, too. The city council adopted the Santa Monica Sustainable City Plan in 1994; almost 30 years later, the city has made measurable progress on projects ranging from retrofitting buildings to embracing renewable energy. The council selects a new mayor every one to two years, ensuring fresh perspectives at the helm. Gleam Davis shared these reflections during her first term as mayor in 2018–2019; she began a second one-year term in December 2022. Davis has served on the city council since 2009.

Below: Situated 16 miles west of Los Angeles on the Pacific coast, Santa Monica is home to 93,000 people.

ANTHONY FLINT: Does Santa Monica's system of having a mayor for two years present a challenge for sustainability efforts, which often are slow to get going—and to pay off? What are the projects that can have the greatest impact through your upcoming term?

GLEAM DAVIS: I don't think it creates much of an impediment to the sustainability agenda. The mayor and the mayor pro tem are members of the entire city council. The city council sets the policy, adopts the budget, and drives the city's policies. Then it's the city manager who does the implementation. Whatever policy direction is given to the city manager is from a vote of the full city council.

On the sustainability front, the big news is we are now part of a group called the Clean Power Alliance, where the default provision for customers is power that is 100 percent sourced from renewables. This is helping us take a big leap toward energy self-sufficiency. People can choose to shift into lower tiers, such as 50 percent renewable, or they can opt out entirely. There are also discount options for low-income families. So far, the opt-out rate is very low.

The not-for-profit Clean Power Alliance provides access to renewable energy from local and regional solar, wind, geothermal, and hydro resources to customers in 32 communities across Los Angeles and Ventura counties.

Another continuing thread is providing mobility choices. We live in a compact city of less than nine square miles, and we have the ability to provide transport options to our residents. We have light rail with three stations, so you can take transit to downtown Santa Monica or downtown LA. Our Big Blue Bus program has a policy of "any ride, any time," so students can get on a bus, show an ID card from any college—a lot of UCLA students ride those lines, along with students from Santa Monica College—and it's free.

AF: The city's overall greening strategy has included a first-of-its-kind zero net energy ordinance for new single-family construction and a commitment that all municipal power needs be met by renewables. But the new $75 million municipal building project has been criticized as too expensive. How can being green be cost-effective?

GD: What's important to know is, we're leasing a fair amount of private property for government offices, at a cost of roughly $10 million a year. We needed to bring employees into a central location, which will save money on leases and will encourage face-to-face and "accidental" meetings that can be so important to communication.

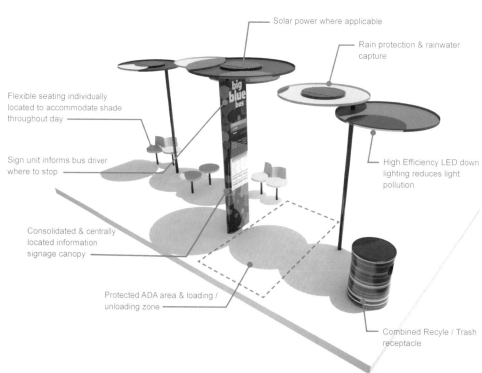

Solar power where applicable

Rain protection & rainwater capture

Flexible seating individually located to accommodate shade throughout day

High Efficiency LED down lighting reduces light pollution

Sign unit informs bus driver where to stop

Consolidated & centrally located information signage canopy

Protected ADA area & loading / unloading zone

Combined Recyle / Trash receptacle

It just made business sense to have everybody under one roof. We'll end up saving money over time, and ultimately the building will pay for itself just on that basis. There will be additional savings over time if the building is energy neutral and has reduced water intake. We've required developers to meet stringent sustainability requirements. If we're going to do that, we need to walk the walk. This project shows that it's possible to build an aggressively sustainable building that will ultimately bring savings. We're trying to be a model, to show that with a little up-front investment, you can have a big impact over time.

AF: How does the Wellbeing Project, which won an award from Bloomberg Philanthropies for its ongoing assessment of constituents' needs, connect to your sustainability efforts? What has it revealed?

GD: We declared ourselves a sustainable city of well-being. How are the people in the community faring? Are they thriving, or are there issues? The Wellbeing Project began as an assessment of youth and how they were doing, and of what we, as a city, can do to try to help. It's really about changing the relationship between local government and people. It's not really a new concept—it goes back, not to be corny, to the Declaration of Independence: life, liberty, and the pursuit of happiness. That doesn't mean people just going out and having a good time; it means people having the ability to thrive. A sense of community can get frayed, whether due to technology or culture. One of the things we do is make sure children enter kindergarten ready to learn. And for our older citizens—are they feeling isolated in their apartments? It's a global movement we're thrilled to be a part of.

In our Wellbeing Microgrants program, if people come up with an idea for a project to build community, we will fund it up to $500. One example was going out and writing down the histories and memories of Spanish-speaking residents in the many parts of the community where English is a second language. Another was a dinner to bring together our Ethiopian and Latino communities. One individual took a vacant lot and created a pop-up play area and space for art. It's about community connectedness.

AF: Another innovative strategy is to impose charges on excess water use to fund energy-efficiency programs in low-income

Above: The Santa Monica City Services building, known as City Hall East, opened its doors on Earth Day 2020. It is the largest municipal building in the United States designed to meet the Living Building Challenge, which requires net zero energy, waste, and water.

Facing page: Santa Monica's Big Blue Bus system provides more than 7.7 million rides annually across a 58-square-mile service area. Today's buses run on natural gas, with plans for an all-electric fleet by 2030, and the system's bus stops incorporate sustainable elements.

It goes back, not to be corny, to the Declaration of Independence: life, liberty, and the pursuit of happiness ... people having the ability to thrive.

homes. What's your long-term view on managing water in what look to be perilous times ahead?

GD: The other thing we've done, which will percolate throughout my term and next, is to work on becoming water self-sufficient. We control a number of wells in the region, but we had contamination in the 1990s, and ultimately reached a multimillion-dollar settlement with the oil companies responsible.

After that, we were getting 80 percent of our water from the Metropolitan Water District—if you saw the movie *Chinatown*,

that's the system that sucks water out of the Colorado River and brings it to LA. But now we've totally flipped that—we're getting 80 percent of our water from our own wells again. This makes us more resilient in case of an earthquake affecting the aqueducts or other disruptive events to water infrastructure, like broken water mains. Pumping water over mountains from the Colorado River also takes a lot of energy. We're making sure our water infrastructure is sound. We're not trying to isolate ourselves, but by using our own wells we will have good clean water for the foreseeable future.

AF: What policies might limit the devastation so sadly seen in the recent wildfires in California?

GD: Luckily Santa Monica was not directly affected by the Woolsey Fire. Our neighbor Malibu was—their emergency operations center was right in the path of the fire, so they came and used ours for fighting the fire, rescuing people, and cleaning up.

We had Santa Monica firefighters on the ground throughout the state under mutual aid. We hosted meetings with FEMA on

Above and facing page: Community projects supported by Civic Wellbeing Partners, a start-up that evolved from the City of Santa Monica's Wellbeing Project.

Santa Monica has broadened its water infrastructure investments with projects including a 1.5-million-gallon stormwater harvesting tank built beneath a courthouse parking lot, a new 1-million-gallon-per-day advanced water recycling facility, expansions to its Arcadia Water Treatment Plant, and restorations to the Olympic Well Field.

displacement and recovery. We have a chief resiliency officer, and she is a steady drumbeat, reminding people that a major natural disaster could happen here. We have promoted the "seven days plan"—does everyone have seven days' worth of water, food, and an emergency radio that doesn't require electricity? We also passed aggressive earthquake requirements, evaluated properties that are most vulnerable, and are now moving to seismically retrofit them.

These things we do in Santa Monica may seem a little aggressive, and cost money, but it's not just about winning awards or patting ourselves on the back for being environmentally progressive; it's so that we'll be able to weather things like fires. People say we're spending money, raising water rates, and that it costs more for energy, but we do it to address the impacts of climate change, and it also means that when there's a natural disaster, we're more resilient.

Below: The Sustainable Water Infrastructure Project (SWIP), an underground treatment facility completed in 2022, is designed to ensure a drought-resilient water supply.

AF: The city's experience with electric scooters—I'm referring to the company that deployed a fleet without asking permission—seemed to show that the transition to a sharing economy coupled with technological innovation can be messy. Is it possible to welcome disruption and maintain order?

GD: We were sort of ground zero for scooters. It was disruptive at first, and we had to make a lot of adjustments. Their philosophy was that it was easier to ask forgiveness than permission. There was some panic, and some people were also using them in a horrible manner. Now we're in a 16-month pilot program, where we selected four dockless mobility operators: Bird, Lime, Lyft, and Jump—which is part of Uber. We created a dynamic cap on the number of devices on the street, so they can't put out as many as they want. We have some policies to address conflicts and safety, and we have issued tickets when necessary.

This is all part of giving our residents lots of mobility choices. The scooter system is designed to give people the option to get out of their cars, whether they're going to downtown LA or walking two blocks to a neighborhood restaurant. We wanted to make sure our more economically diverse communities would have access—it's not just downtown. If you can replace a car with alternative means that include scooters or electric bikes for that first or last mile, that's a big cost savings. We had about 150,000 rides on shared mobility in November 2018. That's pretty amazing for a place with 93,000 people. At the end of the pilot, we'll evaluate everything and figure out where we go from there.

A number of neighboring cities banned scooters outright, but that's not how Santa Monica deals with technology. We're figuring out the best way to manage disruptive technology. Disruption isn't a four-letter word.

Above: A 2021 report on the 2018–2019 dockless scooter and e-bike pilot program found that over the first year of the pilot, 49% of rides displaced drive-alone and ride-hail trips. A second pilot program from July 2021 to March 2023 included nearly 1.5 million shared mobility rides, despite a steep drop in ridership during the pandemic.

A native of California, Gleam Davis holds degrees from USC and Harvard Law School. As corporate counsel for AT&T, she has worked with KIND (Kids in Need of Defense), which represents unaccompanied minors in immigration courts. Before joining AT&T, Davis prosecuted civil rights violations as a trial attorney in the Civil Rights Division of the US Department of Justice and was a partner at the law firm of Mitchell Silberberg & Knupp. Active in the community since moving to Santa Monica in 1986, Davis has been involved with the Santa Monica Planning Commission, Santa Monicans for Renters' Rights, the board of directors of WISE Senior Services, and the Santa Monica Child Care and Early Education Task Force, among many other organizations.

MARVIN REES

At the "Love Local or Lose Local" mural, an example
of Bristol's vibrant street art scene, 2022.

In May 2016, Bristol native Marvin Rees became the first directly elected mayor of Black African-Caribbean heritage to lead a major European city. Rees pledged to make Bristol—a postindustrial city that lies about 100 miles west of London and is home to more than 450,000 people—a "fairer city for all," with a focus on affordable housing, improved transit, health care, and education. The pandemic prompted the extension of his first term to 2021, when he won reelection. When his second term ends in 2024, Bristol will switch to a council committee governing system. Rees chairs Core Cities UK, a group of the 12 largest cities in the UK outside of London; represents the UK on the Commonwealth Local Government Forum; and has advocated for cities at COP climate summits, the United Nations, and New York Climate Week. In this interview, Rees reflects on equity, growth, and immigration, amid an increasingly tumultuous political climate in the United Kingdom.

Below: Formerly a working port, Bristol Harbor has become a popular destination for residents and visitors.

ANTHONY FLINT: One of your campaign billboards indicated you would build 2,000 homes per year once elected. What was behind that promise, and how has it played out?

MARVIN REES: The reason affordable housing became our top priority is because it is one of the single most important policy tools we have for delivering population health, a strong economy, a stable society, and good educational outcomes. We have a housing crisis as many American cities do. We haven't built enough historically, and the private market alone hasn't provided the opportunity to own a stable home. It's been a challenge, in part because we didn't have the organizational machinery in place to bring land forward and get it developed. But we're on track to meet that target—2,000 homes a year, 800 of them affordable—by 2020. There's a whole mix: council houses where we own the land; a social housing associ-ation with rents below market rates; we've got volume builders who, within their schemes, are also required to provide affordable homes; and we are supporting self-build schemes, where com-munities come together to build cohousing on underutilized land. We've hosted the Bristol Housing Festival, which showcased modern methods of construction such as off-site manufacture. We place an emphasis on quality and community; we don't want to just put boxes up and slot people into them.

AF: As you think about sustainable growth and affordable housing, what in your view is the role of land policy, including the taxation of land? Where do you stand on land value capture and a land value tax?

MR: I'm from a public health and journalism background, so I had to have a crash course about how various parts of a city work. Land value is a massive challenge because land has become a com-modity, passing through the hands of several owners, not to be built on but just to make money. We need powers at the local govern-ment level, and the national government needs to take action to change how land is used. There's a huge conversation to be had. In the UK, we think education is a public good. We think the same about health—hence we have a National Health Service. And for social justice and the strength of our economy, we need to reframe how we think about land and housing. If we fail on this, we'll end

Above: Hope Rising, an affordable modular home development built by Zed Pods.

Bristol built more than 2,100 affordable homes between 2016 and 2023, and had over 3,000 additional homes under construction. The city continues to prioritize ex-panding its housing supply, peaking with 2,563 new homes built in 2022–2023. Since 2016, over 12,500 new homes have been built overall.

up with what we've seen across the world—the middle class disappears, and you end up with a bifurcated population and fragile state. This is a crisis.

AF: You have embraced the concept of reinvention for postindustrial cities, which is a big theme of the UK 2070 Commission, a research initiative focused on spatial inequalities in the UK. But how do you encourage growth in your city and others like it in the context of Brexit?

Facing page: Housing surrounds the Bristol Jamia Mosque in the inner suburb of Totterdown.

We're in a postnational world and we can't leave our futures in the hands of national government. . . . Cities are equipped with the political machinery to lead the way.

MR: Brexit is the wrong answer to the right problem. People have been left behind; they've lost hope. They feel that politics has become increasingly distant from them. The other problem Brexit has identified is that people have lost touch with their national story and narrative, and with who they are. Just like in the United States, many want to go back to the 1950s. These are legitimate grievances, but Brexit is not going to solve the problem. Globalization has integrated our communities so we use the same products—there's nothing British about Pizza Hut, right? In many ways we're in a postnational world and we can't leave our futures in the hands of national government. The city level of government is best positioned to deliver, with cities forming international networks to collaborate on shared issues like climate change, immigration, and equity.

AF: Take a moment to explain Bristol's One City Plan, which lays out a vision for where the city will be in 2050 and is shortlisted for the EU's Capital of Innovation prize. How do you balance myriad ideas from constituents with pushing the agenda you have determined is needed?

I'm a physical embodiment of migration, so I think it's disingenuous to say migration is the cause of the world's ills.

Above: Street art overlooks pedestrians in the Stokes Croft neighborhood.

MR: The One City Plan comes from an understanding that what people receive is not determined by government alone; people sit at the intersection of decisions made by the city, universities, and the private sector—and if we want to shape the future, we have to grab hold of that collective impact and get some alignment. We can't wait to see what comes down the railroad tracks. We need to see where we need to be in 2050, and if we want to get there by 2050, what needs to be delivered by 2048 or 2025, and work our way back. It's a living document with shared priorities and real agreement. Anyone in Bristol can pick up a copy of the plan and say, "Right, I see you are doing X by 2050, but I think it should be done by 2025"—carbon neutrality, for example. The One City Plan gives us the raw materials and shows how we can get to common ground.

The plan is based on six themes: Health and Wellbeing, Economy, Homes and Communities, Environment, Learning and Skills, and Connectivity. Each of those stories has a board made up of community members, and they are responsible for updates every year. Every six months we have an event called the City Gathering. For the first one we had 70 or so people come together, and I said to them: Between us we spend £6 billion ($7.3 billion) and employ 70,000 people in the economy. If we align ourselves on a small number of shared priorities, what could we not do? We have incredible power. We're trying to create spaces for people to connect and come up with answers.

AF: As you've been going about your work, you've been the target of extremist and anti-immigration rhetoric. How do you manage being chief executive with a progressive agenda in that kind of climate?

MR: I manage it because I think the whole argument about immigration is, to put it charitably, a mistake, and less charitably, a big lie. Immigration is not the cause of people's problems. I grew up poor and among those often preyed upon. To have members of the British elite running around—and you see something similar in the United States—blaming migrants for the state of the country that they have had all-encompassing power over for centuries— it's a little bit rich. They have created a situation where relatively poor and powerless people are blaming other poor and powerless people for the state we are in. It's not difficult for me because I want to be in a place where I can say what I really think. I'm a mixed-race man. My dad came from Jamaica; my mum's English heritage goes back in Bristol for a very long time. My granddad was from South Wales and before that Ireland. I'm a physical embodiment of migration, so I think it's disingenuous to say migration is the cause of the world's ills.

Another problem is that the migration discussion is being shaped by national governments. That's the wrong way around. What we need is for national governments to start talking to cities and asking what cities need. Cities are more inclined to look at migration as an asset in terms of our connectivity to world markets. Our Asian, African, or Eastern European populations connect us to international opportunities. National governments are using abstract numbers and talking about how many more people to

City leaders have updated the One City Plan several times since 2019, incorporating feedback from participants, partners, and community members. The six themes have evolved into: Children and Young People, Economy and Skills, Environment, Health and Wellbeing, Homes and Communities, and Transport.

In 2020, Black Lives Matter protesters toppled a statue of a slave trader and threw it into Bristol Harbor. Rees earned national and international attention for his response to that event, facilitating citywide conversations about race, power, and place.

In 1971, the population of Bristol was 406,000, and 97.7% of residents were white. That demographic shifted over the decades: in 2021, according to the Office of National Statistics, the population of 472,467 was 81% white, 6.7% Asian, 5.8% Black, and 4.5% mixed race, with 2% of residents identifying with another ethnic group.

Above: As a coastal city, Bristol is particularly vulnerable to climate change. The city estimates that thousands of properties are already at risk from tidal and surface flooding, lending urgency to its efforts to build resilience.

Facing page: Some 200 acres of new wetland habitat are being created in the north of Bristol, as part of a 10.5-mile flood defense project along the Severn Estuary.

let in. And that's completely different from the conversation we need to have.

AF: Last but certainly not least, what is your vision for how cities like Bristol can contribute to combating climate change, while also preparing for its inevitable impacts?

MR: We absolutely recognize it as a crisis with very real consequences. Increased flood risk, more extreme temperatures, desertification—we'll end up with more rural-urban migration, and a source of conflict leading to more crises. For cities, the climate emergency will be inseparable from the global migration emergency. Cities have to be in the driving seat for a number of reasons. One is about political will. Certainly in the United States, your federal government seems to have no political will, but we've seen American mayors stepping up to lead when the federal

government withdraws. Cities are more inclined to look in terms of interdependencies, whereas the national government is more occupied with boundaries. Cities are equipped with the political machinery to lead the way.

Born in Bristol to an English-Welsh mother and Jamaican father, Marvin Rees grew up in the city's public housing, in a climate of both economic and racial tension. He studied economic history and politics at Swansea University, then global development at Eastern University in Pennsylvania and at the Yale World Fellows global leadership program, before returning to the UK. Rees worked in public health and journalism before seeking office, having graduated from Operation Black Vote and the Labour Party's Future Candidates Programme.

MARTY WALSH

Mayor's office, Boston City Hall, 2020.

During his two-term tenure as Boston's 54th mayor, Martin J. Walsh focused on schools, affordable housing, and immigration, among many other issues. He also became an international leader in confronting climate change and building resilience, hosting a major climate summit in 2018 and chairing Climate Mayors, a coalition of US mayors committed to working on renewable energy and other strategies. While in office, Walsh pledged to make Boston carbon neutral by 2050; he also led Imagine Boston 2030, the first citywide comprehensive plan in half a century, and the Resilient Boston Harbor initiative. After leading the city for seven years, Walsh was named US secretary of labor. He held that position for two years before resigning to become head of the National Hockey League players' union, a role he has described as "more like being the mayor." In this interview, he reflected on the challenges and opportunities facing local leaders in the midst of the unfolding climate crisis.

Between 2014 and 2021, Walsh served as both chair and cochair of Climate Mayors, a coalition of 750 current and former US mayors across 48 states who have committed to climate leadership and upholding the Paris Climate Agreement.

ANTHONY FLINT: You have been one of the most active mayors in the nation on the pressing issue of climate change. Tell us about your recent efforts to coordinate action—and how you feel about all this work being done at the local level in the absence of a federal initiative?

MARTY WALSH: We hosted our first International Mayors Climate Summit in 2018, and we've been working with mayors across America. I was elected as the North American cochair for C40 prior to President Trump pulling out of the Paris Climate Agreement. We've been working with Mayor Garcetti in Los Angeles and other mayors to make sure that cities recommit themselves to the agreement. This is such an important issue for the country and for Boston, and it's essential to have engagement and leadership. It's unfortunate that we haven't had a federal partner for the last few years. But we're going to continue to take on the challenges and to think about the next generation. I'm hoping that ultimately we will have a federal partner, and when that time comes, we won't be starting at zero.

AF: Turning first to mitigation: what are the most important ways that cities can help reduce carbon emissions? Should cities require the retrofitting of older buildings, for example, to make them more energy efficient?

MW: We have a program called Renew Boston Trust, identifying energy savings in city-owned buildings. It's important to start in our own backyard. We have 14 buildings underway for retrofits—libraries, community centers, police and fire stations. Secondly, we're looking at electrifying some of our vehicles. The third piece is making sure all new construction is built to higher performance standards, with fewer carbon emissions. Ultimately, as we think about reducing carbon emissions, we are looking at 85,000 buildings in our city. If we want to hit net zero carbon by 2050, we'll have to retrofit those buildings, large and small. Then there's transportation—getting our transportation system to be cleaner and greener. Even if we eventually have a strong national policy, it's the cities, ultimately, that will have to carry out the reductions.

Launched in 2017, Renew Boston Trust expanded its scope in 2022 to include an energy audit of all exterior city-owned lighting; energy-efficiency and carbon-reduction upgrades within the facilities of the Boston Public Schools; and expansion of the city's fleet of solar arrays on municipal properties.

Below: Pedestrians near Boston's Old State House and Financial District.

AF: Even if we were to stop all carbon emissions tomorrow, the planet will still have to manage significant sea-level rise, flooding, volatile weather, wildfires, and more, because of inexorably rising temperatures. What are the most promising efforts in Boston and around the country in building resilience?

Resilient Boston Harbor maps out a strategic vision to make the city's 47 miles of coastline less vulnerable to flooding and sea-level rise, and more accessible to the public. The plan focuses on adapted infrastructure, protective waterfront parks, and elevated harborwalks.

MW: For Boston and East Coast cities and oceanfront property, our Resilient Boston Harbor plan lays out some good strategies. We have 47 miles of shoreline, and rivers that run through and border our city. In terms of protecting people in major flooding events, we've looked at the impact of Superstorm Sandy, and at what happened in Houston due to Hurricane Harvey in 2017. We have one big plan for the harbor, but there are other neighborhoods where we need to make sure we're prepared. We're doing planning studies in all of these areas under the Climate Ready Boston initiative to deal with sea-level rise.

The Climate Ready Boston initiative includes studies of the city's five coastal neighborhoods, delving into heat resilience, green infrastructure, natural hazard mitigation, open space and urban forest planning, and wetlands protection.

Climate change is a public safety matter. It's about quality of life and the future of our city. In the past, mayors have focused on economic development and transportation and education. Today, climate change, resilience, and preparedness are part of the conversation in ways they weren't 25 years ago.

AF: The Dutch and others have developed innovative ways of working with nature through blue and green infrastructure, incorporating water elements and vegetative areas. Are you also a fan of this approach?

Right: Boston University's Center for Computing and Data Sciences, the city's largest fossil fuel-free building, relies on deep thermal wells for heating and cooling.

MW: Resilient Boston Harbor is really a green infrastructure plan. One project that speaks to that is Martin's Park, named for Martin Richard. We raised parts of the park to prevent flood pathways, and installed mini piles and vegetated beds reinforced with stone to prevent erosion at higher tides. We're looking at doing something like that throughout the inner harbor. We're spending $2 million at Joe Moakley Park, in South Boston, where major flood pathways to several neighborhoods start. We're trying to cut back on as much flood-related property damage and disruption of people's lives as possible. Berms and other barriers can help keep the water out, but there are also opportunities to let the water through and not let it build up in a major storm event.

AF: In addition to new taxes that have been proposed, would you support a land value capture arrangement where the private sector contributes more to these kinds of massive public investments?

MW: On top of private investment—which we're going to need more of—we are working with philanthropic organizations to see if they can fund these kinds of projects. In our budget this year, we're dedicating 10 percent in capital budget to resilience. We're also looking at taking some dedicated revenue and putting it into resilience. For example, we raised fines and penalties for parking violations. That will go right back into transportation and

Above: The climate readiness plan for Charlestown, Boston's oldest neighborhood, includes waterfront features such as elevated pathways and floating wetlands meant to provide access to open space and protection from flooding.

Boston is asking developers who wish to build in its low-lying Marine Industrial Park to contribute to a Climate Resiliency Fund, which will help the city finance a seawall and other flood defenses that, in turn, will directly protect the developers' properties.

Part of our job is to govern in the present day, and manage all the day-to-day operations; but our job is also to lay down the foundation of what our city will look like in the future.

resilience, including things like raising streets up. That's a start. Over time, we'll dedicate more of our budget to this. At some point, hopefully, the federal government will invest in preventative measures like these, instead of paying millions and millions for disaster relief. Rather than coming in after an event and a tragedy happens, I hope that they will make investments on the front end.

AF: Given projections that large swaths of Boston will be underwater later this century, can you reflect on a personal level about this threat to the city you currently lead? How would you inspire more urgency to address this problem?

MW: That's our job. Part of our job is to govern in the present day, and manage all the day-to-day operations; but our job is also to lay down the foundation of what our city will look like in the future. The infrastructure that we build out will be here for the next 50 to 60 years. The Resilient Boston Harbor plan is designed to deal with sea-level rise 40 or 50 years from now. We're building all of that with the expectation of preserving and protecting the residents of the city. I would hope that when I'm not here as mayor anymore, the next mayor will come in and will want to invest as well. This is the legacy of the city—I wouldn't say it's necessarily my legacy—for residents to look back years from now and be grateful for the investments and the time that leaders took in 2017 and 2018 and 2019.

I don't think as a country we're where we need to be. The Dutch and other European countries are farther ahead. So we're playing catch-up. And we're not waiting for the next generation to try to solve this problem.

Facing page: Martin's Park, named for the youngest victim of the 2013 Boston Marathon bombing, is a climate-resilient space designed to help protect the city from sea-level rise.

Born to Irish immigrant parents and raised in the working-class Boston neighborhood of Dorchester, Marty Walsh began his career as a construction worker and union leader. In 1997, he was elected to represent Dorchester in the Massachusetts House of Representatives, where he served until becoming mayor in 2014. Walsh has been candid about his challenges, which include surviving cancer as a child and dropping out of college; he earned a BA from Boston College in 2009, at age 42. He served as US Secretary of Labor from 2021 to 2023 and is executive director of the National Hockey League Players' Association.

KOSTAS BA

KOYANNIS

Meeting with a resident of Anafiotika, a neighborhood below the Acropolis, 2019.

When this interview was conducted in early 2020, Kostas Bakoyannis—
the youngest person ever to be elected mayor in Athens—had been
in office less than a year, and the COVID crisis was just emerging.
Since then, Greece has rebounded from the pandemic and is focused
on ensuring its long-term sustainable recovery. Athens has also experienced a record inflow of investments and tourism, facilitating
its "twin transformation" into a green and digital city. However, old
wounds from more than a decade of inertia and new challenges
stemming from rising costs of living, climate change, and security
and migration issues are affecting the livelihoods and well-being
of Athenians, who voted a new mayor into office in late 2023.

Below: Athens Street, looking toward the Acropolis.

ANTHONY FLINT: You have said that you are focused not on grand projects, but on day-to-day quality of life in a city trying to make a comeback in a more incremental fashion. What are your reflections on your successful campaign and the experience thus far of being at the helm of local government?

KOSTAS BAKOYANNIS: In any campaign, it's about the message—not the messenger. Elections in the past in Greece have been about candidates higher up talking down to the people. I took a different approach and started walking out in the neighborhoods. I listened with care and found that the people want our city to build its self-confidence and be optimistic again. Now we are reinventing city services and reinventing the city.

Athens holds three records: it has the least urban green per capita in Europe, the most asphalt, and its houses have the most square meters in relation to lot size. We want to reclaim public space, and especially reclaim space from the automobile. We've been studying traffic circulation—we plan to close parts of the city center to cars, and we're creating an archaeological walkway around the city.

All in all, I'm living my dream. I'm giving it my all. I've been in local government for 10 years; higher office doesn't compare. One day, when I first began my journey in local government, I was depressed, thinking we are a failure, and then I walked out and saw a playground we had just opened. This job is not about resolving the conflict between North and South Korea. It's about real, tangible, incremental change—improving the quality of life.

AF: Athens has been vexed over the years by the problem of vacant buildings and storefronts, graffiti, homelessness, and a general image of being dark and dirty. Can you tell us about your plans to clean things up?

KB: There was a very good article in an international magazine about the Greek economy, but at the top there was a photo of Athens, with two homeless people sleeping in front of closed stores that were full of graffiti. This is our challenge. We're in a global race to attract talent, technology, and investment, and Athens is changing day by day. We have adopted the "broken windows" theory of social behavior and are coordinating with

The term "broken windows" refers to a controversial theory in criminology that visible signs of crime, vandalism, and decay in urban environments invite more of the same.

This job is not about resolving the conflict between North and South Korea. It's about real, tangible, incremental change— improving the quality of life.

Facing page: The city is prioritizing graffiti removal and other basic improvements in long-neglected public spaces.

The Adopt Your City program invites Athenians to sponsor public areas such as parks, playgrounds, and schools to help contribute to their care and revitalization.

In 2022, the City of Athens opened the first supervised spaces for drug use in Greece.

the police. We run campaigns and have special equipment to clean up graffiti. We've started a program called Adopt Your City, and have created public-private partnerships that are already bearing fruit. We are asking people who care and love the city to come help us. Regarding drugs, reforms have been made. The Parliament recently passed a measure on supervised spaces for drug use—we haven't operated one yet, but we are preparing to make it mobile, so it doesn't stay too long in any one neighborhood. Local government will be able to operate such spaces. We are reclaiming public space, like Omonia Square, a city landmark— it's going to be a symbol. It's not just public works; we are creating more of a product, an experience.

AF: As part of that effort, you attracted controversy for clearing out squatters in the neighborhood of Exarchia, an effort that included dawn raids and relocating refugees and undocumented immigrants. How do you fulfill your campaign promise to restore law and order and curtail illegal immigration, while still being sensitive to the human lives at stake?

KB: Here is an example: An individual calling himself Fidel was running a hostel in a school, occupying it, and charging money. We securely moved the children to take advantage of social service provisions. Greek media have a thing about Exarchia. It becomes a political weapon for one side or the other. I don't look at it that way. We have 129 neighborhoods, and Exarchia is a neighborhood with its own issues. Much of what we do has to do with persisting and insisting—it's a question of who will get tired first. We will not get tired first.

Above: Bakoyannis has drawn a distinction between street art and graffiti: "We want a vibrant, creative city.... We say yes to color and art, and no to the abuse of public space."

On the subject of pluralism, we're the canary in the coal mine. We survived the economic crisis, and we're stronger today than we've been for the past 10 years. We have more depth to our democracy, stronger institutions. We isolated extremists. We confronted the Fascist Nazi party Golden Dawn. We went to neighborhoods where they were doing well, but we didn't wag our fingers and tell people they were bad for voting for Golden Dawn. Instead, we said: we can provide better solutions to the problems you face.

Athens is a Greek city, a capital city, and a center for Greeks around the world. Having said that, Athens is changing and evolving. I remember seeing a young woman who was Black in a parade, and she was proudly holding the flag—I think what she was saying was, "I'm as Greek as you are." We want to make sure everyone living in the city has the same rights and obligations.

AF: What are the most important elements of your plans to help Athens combat climate change—and prepare for its inevitable impacts in the years ahead?

KB: Think different! It is all about working bottom up. In terms of public policy, cities are true laboratories of innovation. Nation-states are failing—there's so much partisanship, and bureaucracies that cannot handle real problems; cities are closer to the citizen. We are proud to be part of C40. Athens has developed a policy for sustainability and resilience. Among other things, we are working on ambitious but realistic interventions to liberate public space, multiplying green space and creating car-free zones. For us, climate change is not a theory or an abstraction. It is a real and present danger that we can't just sweep under the rug. It demands concrete responses.

AF: You recently had the opportunity to return to Cambridge and Harvard. What level of interest did you find in the future of Athens? Are there things you have learned from American cities, and what can the United States learn from you?

KB: I was enthused and heartened by the level of interest, and was proud to represent a city with a glorious past and a promising, bright future. We may live on different sides of the Atlantic, and in very different cities, but we face similar challenges as urban centers evolve. It is always great to share experiences and learning moments, like policies to further climate resilience. And of course, battling social inequalities is at the top of all of our agendas.

In addition to implementing an extensive urban greening plan and resilience capacity-building programs, Athens appointed Europe's first chief heat officer, Eleni Myrivili, in 2021. The appointment was part of a program led by the Arsht-Rockefeller Foundation Resilience Center and the Extreme Heat Resilience Alliance that saw a chief heat officer named on each continent.

Below: Built in 1846, Omonia Square—an iconic public space— was restored in 2020. Integrating green elements such as LED lighting and permeable pavement, the landmark is intended to double as "an oasis of cool" as urban heat increases.

Kostas Bakoyannis was elected mayor of Athens in 2019 at the age of 41. The son of two prominent Greek politicians, he previously served as mayor of the town of Karpenissi and regional governor of Central Greece. Bakoyannis earned his undergraduate degree at Brown University, a master's degree at Harvard's Kennedy School of Government, and a PhD in political science and international relations at the University of Oxford. He has served in the Greek Ministry of Foreign Affairs, the European Parliament, and the World Bank, and holds positions with the European Council on Foreign Relations and the United Nations Sustainable Development Solutions Network.

Next page: In 2023, a record-setting wildfire broke out on Mount Parnitha, a national park within sight of the Parthenon. Bakoyannis has called climate change—which is causing more severe and more frequent wildfires—a "real and present danger."

LIBBY SCHA

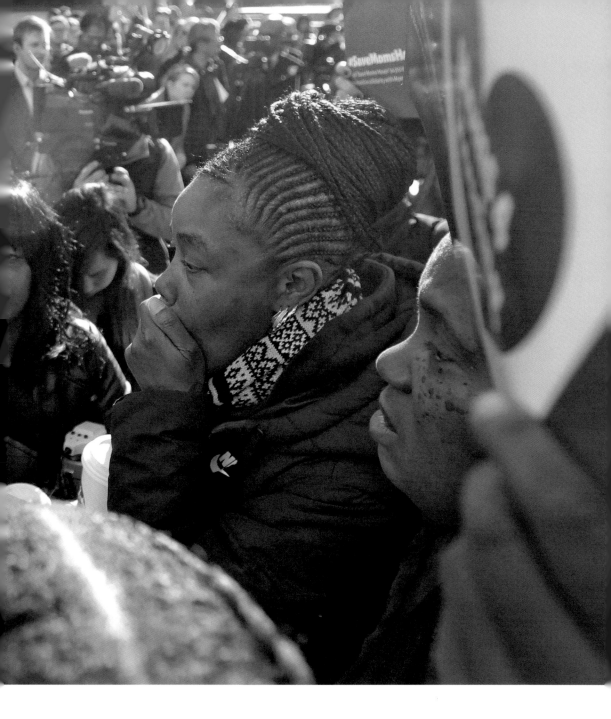

Consulting with the founding members of Moms 4 Housing, an activist organization that inspired state housing reform, at a protest at Oakland City Hall, 2020.

During Libby Schaaf's two terms as mayor of Oakland, the city underwent an economic revitalization and building boom and cut gun violence nearly in half. Schaaf worked to stabilize rents, decrease evictions, and address homelessness, and was appointed to California's first Council of Regional Homeless Advisors in 2019. She created Oakland's first Department of Transportation, prioritizing projects in underserved neighborhoods and making headway on the city's decades-old infrastructure backlog. She also launched Oakland Promise, a cradle-to-career initiative intended to improve educational outcomes for local students by providing scholarships and other support. In the spring of 2020, shortly before giving this interview, Schaaf responded to the coronavirus crisis with a successful lockdown, reconfiguring streets for pedestrian uses and working with the county and state to make vacant hotel rooms available for unhoused people. After leaving office in January 2023, she became interim executive director of Emerge California, a nonprofit that recruits and trains women leaders. She teaches at the Goldman Public Policy School at UC Berkeley.

ANTHONY FLINT: How did the pandemic unfold for you, and how has your job changed since this all began?

LIBBY SCHAAF: We were one of the very first places to have to confront this crisis. I remember I got the call from Governor Gavin Newsom; we were being asked to allow the *Grand Princess* cruise ship to disembark. We're all members of an interdependent human community that must put people over profits, and must put health first, and we must model generosity and the right values in moments of crisis. So I accepted the request. We made sure to constantly remind people of the historical context of environmental racism in West Oakland—the city's port area—and we also pushed our federal and state partners to take extra measures to ensure community and worker safety during the operation. It was heartening to be able to provide that safe harbor, but for me,

as mayor, it was an unusual entry into the COVID crisis because it focused my attention externally, toward these outsiders that needed help from my city; then I needed to quickly pivot internally as we began to see cases unfold in our own community.

I have only been to City Hall once in the last nine weeks. So my life has changed tremendously on a personal level. I've learned that I'm really bad at cutting my son's hair. I've learned that my husband is an amazing cook. And I've learned that our Slow Streets in Oakland are an amazing comfort—I got to enjoy the first one in my neighborhood recently.

A quarter of my workforce in Oakland has lost their jobs, and that is such a sobering reality. I have shifted a tremendous amount of my time and energy to communications and crisis management.

Above: The Port of Oakland. In March 2020, the *Grand Princess* cruise ship was sent to the port after reporting one of the nation's first COVID outbreaks.

Next page: More than 2,400 passengers and 500 crew members disembarked from the *Grand Princess* in Oakland; among those who agreed to be tested, 122 cases were confirmed.

Californians lost more than 2.7 million jobs in the first two months of the pandemic. The state reported that it had recovered the jobs in November 2022.

But I also want Oakland to be the "silver lining" city. I want us to exploit every opportunity in this crisis to make lasting structural change that needs to be made and needed to be made before the crisis. COVID is only exacerbating things like structural racism, like economic disparity, and we have an opportunity to not just respond in the moment, but to make enduring changes and to take advantage of a new level of national awareness. The problem of health disparities by race is finally getting national attention. Let's take that awareness, let's take what is hopefully an elevation of political will, and pass some laws so that we don't see these kinds of disparities again.

Oakland was one of the first cities in the country to create a Slow Streets program, identifying 74 miles of streets that could be closed to cars. City leaders also launched the Essential Places initiative, which supported safe access to grocery stores, food distribution sites, and COVID testing locations. The city reopened the streets to vehicles in early 2022, but is incorporating the Slow Streets philosophy into a city-wide expansion of neighborhood bike routes.

AF: You've received a lot of attention for building on the Slow Streets program and for extending closures and other measures to accommodate social distancing and encourage biking, walking, and using scooters. Do you think this might be a turning point for the public realm in cities everywhere?

LS: Without question, this is a turning point. I really want to shout out to my director of the Department of Transportation, Ryan Russo, who saw the opportunity to repurpose analysis that had already been done to create our bike plan. He knew in an instant which 74 miles—10 percent—of our roadways were eligible to be shut to through traffic and used as Slow Streets. So again, recognizing that you can repurpose work you've already done in a crisis is brilliant.

We're experimenting, and government doesn't do that enough.

I know that people are worried about the future of public transit. People are worried—are we going to go back to the car culture where we have solo drivers in their isolated pods spewing pollution and emissions and everyone stressing out because we're all stuck in traffic jams? We cannot go back to that. Slow Streets are a place to give our residents a mental health respite by being outdoors; they offer people a safe, convenient place in their own

neighborhoods to exercise, and to send their kids out on their scooters and roller skates to blow some steam off in a way that is socially distant.

But it's a reminder that the public right-of-way belongs to the public. It's not just for cars. And we've been underutilizing this precious asset in so many ways. It has been truly uplifting to see the joy of our residents enjoying those Slow Streets. We're experimenting, and government doesn't do that enough. And that's once again a silver lining of this pandemic—that people are trying things and the tolerance for risk-taking from our public is much higher; they realize this is a crisis and we've got to do things differently. We've got to test things out, but we're doing it in a very responsive way. We are getting continual feedback from the public about what they like and what they don't like. We're shifting, we're evolving, but I can promise you that much of this is going to remain once the health crisis is over. People are loving it.

AF: Mass transit is really the lifeblood of cities, and its fate is uncertain. How can mobility policies adapt to this new reality?

LS: Well, I want to start with our bus system because they made some quick changes that have been wonderful. And again, these may be things that we're never going to go back on. They stopped

Above: Masked customers and vendors at the Old Oakland Farmers' Market in April 2021.

charging fares. Boom. That allows them to board people from the back doors of the bus as well as the front; people are on and off faster. They don't have to go by the driver. It's less stress on the drivers, faster operations, less touching and physical exchange. There are ways that we can kind of socially distance within public transit. Now, that's going to be a lot harder on a BART car. I used to ride BART several times a week and I joked about getting close to my constituents—because you definitely were pressed up against your fellow man! That is not something that's going to make sense for some time.

But rather than get people back in their cars, we're excited about trying to accelerate the use of electric bikes and scooters. Oakland used to have the nickname "Oaksterdam," mostly because of our embracing of cannabis. But let's be Oaksterdam for bike riding. It's a healthier way of getting around. It's safer. And that is a pivot that we can make. All I know is that we cannot afford to get back into our cars again. That is not an option.

AF: All the economic disruption at a macro level is just staggering, but in terms of cities and downtowns, it's hit some of the basic building blocks—retail, restaurants, office space, residential. How is this likely to alter our urban future?

Below: Pedestrians, bicyclists, and drivers share space at 10th and Broadway.

LS: We've seen construction continue; people are continuing to pull permits. The Bay Area had a significant housing crisis before COVID and arguably we will have an even worse one after COVID, particularly for affordable housing and addressing our homelessness problems. So I have been encouraged to see development continue. . . .

We are already working with our business districts to see if they might want us to close streets off so that restaurants and shops can spread out onto the sidewalk and into the roadway. We know that this virus is far more deadly in interior spaces—particularly when people remain in a small interior space for long periods of time—so anything you do outdoors is going to be much safer. So why not create more of a marketplace atmosphere that could make some of these commercial areas even more attractive?

Cities are not going away. Just because you have to socially distance does not mean you can't do that in a small apartment with someone above and below you and next to you. Cities are efficient. They allow us to deliver services more rapidly. Sprawl is not a healthy response. Smart density and the agility and creativity of cities is what's going to allow us to not just get through this health crisis, but to emerge with a more equitable, healthy environment. Look at the lessons we've learned about how our planet is benefiting from reduced car emissions. We've got to keep these lessons and not just go back to business as usual.

I know it seems like a funny time to sound optimistic. This is a horrid tragedy. Like many people, I have lost loved ones to this disease, and the severity of it is not to be underestimated in any way. And yet we have to see these opportunities. We have to seize them and that is our challenge right now. Anyone who's mayor of a major city has to be a bit on the optimistic side. That's what keeps us going, particularly during times like these.

Global greenhouse gas emissions decreased 4.6% in 2020, according to the International Monetary Fund Climate Change Indicators Dashboard—but rebounded in 2021, increasing by 6.4% and setting a new record.

Above: The Flex Streets Initiative, launched in June 2020, allowed businesses to expand their outdoor operations, resulting in 140 sidewalk cafes and parklets, 13 street closures, 14 permitted private spaces, and more than 60 mobile food truck permits over the next two years. The city council voted to make the program permanent in 2022.

Libby Schaaf was born and raised in Oakland. She earned her BA in political science from Rollins College in Florida and her JD from Loyola Law School in Los Angeles. Schaaf built her career in Oakland, working as an attorney before joining a local non-profit organization, where she created and ran a volunteer program for the Oakland schools. She served as a legislative aide to the Oakland City Council president and as a special assistant to Mayor Jerry Brown, then as head of public affairs for the Port of Oakland. After serving as economic policy advisor to the city council for a year, she was elected to a council seat in 2010. Schaaf served as mayor of Oakland from 2015 to 2023. She holds senior fellowships with Mayors for a Guaranteed Income and the Harvard Graduate School of Education.

MURIEL BO

Surveying the freshly painted Black Lives Matter mural from the roof of the Hay Adams Hotel, 2020.

WSER

Muriel Bowser vaulted to national prominence in 2020 as a leading voice in both the coronavirus pandemic and the movement for racial justice. Bowser had been elected mayor of Washington, DC, in 2014, then reelected in 2018. She would win office again in 2022 with 77 percent of the vote, becoming the second mayor in the city's history elected to three consecutive terms. A vocal proponent of DC statehood, Bowser serves in a unique capacity, functioning as a governor and county executive as well as mayor. Since taking office, Bowser has sought to speed up affordable housing production in the District, which is home to 706,000 people across 68 square miles and has a budget of $16 billion. She has also worked to diversify the local economy, address rising crime rates, revitalize the downtown corridor, and invest in programs and policies that support families.

ANTHONY FLINT: You came into office in early 2015. Was there anything that could have possibly prepared you for 2020—and how do you see the rest of this tumultuous year playing out? Are you confident about the management of the coronavirus?

MURIEL BOWSER: As a global city, we are constantly preparing for a wide range of shocks and stresses. However, COVID is clearly an unprecedented event that has required an unprecedented response. Our residents and businesses have made tremendous sacrifices for the health and safety of our community. As a district, we are fortunate that we went into this crisis strong. That enabled us to immediately begin putting in place resources to protect and support residents—from delivering meals to seniors, to creating free grocery distribution sites for anyone in need, to setting up free testing sites across the city and quickly hiring hundreds of contact tracers. Since the beginning of this emergency, we have been very focused on following the science, listening to the experts, and keeping our community informed. I expect

that to continue until we get to the other side. Overall, though, I am very proud of how Washingtonians have responded to this challenge.

AF: What did the Black Lives Matter mural on 16th Street in front of the White House—duplicated in many other cities—tell you about the dynamics of public space and social change?

MB: I decided to create Black Lives Matter Plaza when peaceful protests against systemic racism were met with tear gas, and federal helicopters, and soldiers in camouflage roaming our local streets. So we created a place where Americans could come together for protest and redress, for strategizing and healing. Americans nationwide have taken to the streets to demand change. Whether it's through protest or art, or a combination of protest and art, people are using public space to send a clear message that Black lives matter, that Black humanity matters, and that we need to have this reckoning and fix the broken systems that, for too long, have perpetuated racism and injustice.

Above: Visitors to the Lincoln Memorial during the summer of 2020.

Next page: On June 1, 2020, federal officers fired tear gas and rubber bullets into a crowd of people who were peacefully protesting racial injustice and police brutality near Lafayette Square. Four days later, Bowser unveiled her own form of protest: she had authorized artists and city workers to paint "Black Lives Matter" in 48-foot-tall yellow letters spanning two blocks of 16th Street, and christened the area Black Lives Matter Plaza. DC's city council voted to make the mural a permanent installation in October 2020.

Above: More than 23,000 people, primarily immigrants and Black residents, were displaced from thriving neighborhoods in South-west DC (left) by demolition and urban renewal (right) in the 1950s.

By September 2023, the city was 85% of the way to its new housing target, with 30,620 total units built, and 67% of the way to its affordable housing target, with 8,086 new affordable units built.

AF: In 2019, you set a goal of creating 36,000 new housing units—12,000 of them affordable—by 2025. What are the key things that need to happen to create more housing options in Washington?

MB: When I came into office, we more than doubled our annual investment in DC's Housing Production Trust Fund, to $100 million per year. That's more per capita than any other jurisdiction. And we aren't just investing—we've been getting that money out the door and into projects that are producing and preserving thousands of affordable homes across our city. But we have to do more. As you highlighted, we have a big goal in DC: to build 36,000 new homes by 2025, with at least a third of those units affordable. Last year, we became the first city in the nation to set affordable housing targets by neighborhood. When we announced those targets, we also hosted community conversations across DC to discuss with residents the ongoing legacy of redlining and other discriminatory practices and how we can work together to do better. Some of the things we are doing to make this happen are a tax abatement for high-need areas, changes to our inclusionary zoning program, and continuing those big—and strategic—investments in our Housing Production Trust Fund.

AF: If the city's economy recovers from the pandemic, Washington is likely to continue its urban renaissance success story. What policies do you have in place to address gentrification and displacement, both residential and commercial?

MB: Washington will bounce back from this pandemic. We are still home to more than 700,000 Washingtonians who are resilient, creative, and focused on helping their neighbors get through this; and knowing that, I know we will come back.

Going into the public health emergency, we were already very focused on building a more inclusive city and making sure the benefits of our prosperity were felt by more Washingtonians. This pandemic has only amplified the importance of our equity efforts. And as we move forward with our response and recovery, we are still focused on how we advance our goals around housing, jobs, health care, and more. We're still investing more than $100 million in affordable housing. We're moving forward with our strategic plan to end homelessness and we're opening new, more dignified shelters across our city. Our homeownership programs continue. We're looking across the housing continuum to see how we can help more Washingtonians stay and build their futures in DC.

Below: The Triumph, built in 2018, provides short-term housing for vulnerable families.

Cities are incubators for innovation, and while we don't always have the same challenges . . . we are constantly learning from each other.

And we are also supporting our small businesses and local entrepreneurs. For example, we recently announced a new equity inclusion strategy that will increase access to development opportunities for organizations that are owned or majority controlled by individuals who are part of a socially disadvantaged population.

AF: What kind of importance do you ascribe to the planning office of your city, and by extension, who is doing a good job in the practice of planning in other cities?

MB: It's critical that we not only plan for the long-term growth of DC, but also make sure our growth reflects the values of an inclusive and vibrant city. My Office of Planning plays a crucial role in advancing our housing goals and helping us build a city that works for Washingtonians of all backgrounds and income levels. Because the planning office can provide policy analysis, long-term thinking, and community outreach, as well as the implementation needs around zoning and land use, I consider them one of our housing agencies. They work alongside the traditional housing department, public housing authority, and housing finance agency to address housing and affordability.

Across the country, there are so many fantastic things happening at the local level—cities and towns are building innovative solutions to match their unique needs, from Los Angeles to Gary, Indiana, to Boston. Cities are incubators for innovation, and while we don't always have the same challenges—for example, some cities have a lot of people and not enough housing, others have a lot of housing and not enough people—we are constantly learning from each other.

AF: What can cities do now to address the climate crisis, which grinds on, though eclipsed by the other emergencies that have been more front and center?

MB: Environmental justice must be part of the larger conversation we are having as a nation right now. We know, for example, that the harm caused by manmade climate change disproportionately affects communities of color. Additionally, when we look at the disproportionate impact that COVID is having on Black Americans, that is directly tied to the work we must do to build healthier and more resilient communities. This is all a conversation about equity and justice. In DC, we have several programs, like Solar for All, that are focused on fighting climate change while also addressing inequality and other disparities. We don't have to silo these issues; we can and we must focus on all of it.

Above: City leaders are taking steps to combat displacement in gentrifying areas like Logan Circle.

Solar For All, launched in 2016, provides access to solar equipment for low- and moderate-income families.

Born and raised in Washington, DC, Muriel Bowser studied history at Pittsburgh's Chatham College and received her master's degree in public policy from American University. She entered politics in 2004 as an advisory neighborhood commissioner in the Riggs Park neighborhood, then became Ward 4 councilmember in a 2007 special election, winning reelection in 2008 and 2012 before becoming mayor in 2014.

KATE
GALLEGO

With students participating in Camp O'Connor, a civics and leadership development program at the Sandra Day O'Connor Institute for American Democracy in Phoenix, 2023.

Phoenix is the fifth-largest city in the United States and one of the nation's fastest-growing metropolises. For Mayor Kate Gallego—the second elected female mayor in the city's history—navigating that growth has meant prioritizing economic diversity, investments in infrastructure, and sustainability in an area facing extreme heat and worsening water shortages. As vice chair of the Climate Mayors coalition and C40 Cities, Mayor Gallego's goal is to make Phoenix the most sustainable desert city on the planet. Gallego was elected in 2018 to complete the term of a mayor who was heading to Congress, then reelected in 2020. As a member of the Phoenix City Council, she led the campaign to pass Transportation 2050, a $32 billion, voter-approved plan that represented the country's largest local government commitment to transportation infrastructure when it passed in 2015. She has also led efforts on criminal justice reform and ensuring equal pay for equal work. Gallego sat for this interview just a few days after her reelection.

Facing page: The desert city of Phoenix is home to 1.6 million people, with the metro area topping 5 million for the first time in 2023.

ANTHONY FLINT: Congratulations on your reelection. What issues do you think motivated voters most in these tumultuous times?

KATE GALLEGO: Voters were looking for candidates who would deliver on real data-driven leadership and science-based decision-making. I come to this job with a background in economic development and an undergraduate environmental degree. My chemistry professor once told us that the more chemistry you take, the less likely you are to move up in electoral politics. But I think 2020 may have been a different kind of year; in this election, science mattered. Arizona voters wanted leadership that would take COVID seriously, as well as challenges like climate change and economic recovery.

For younger voters in particular, climate change was a very important issue. I ran for office as my community faced its hottest summer on record. In some communities, climate change may be a future problem, but in Phoenix, it's facing us right now.

Previous page: Infrared photo-graphy reveals searing heat in Phoenix, which experienced a record 31 straight days of temperatures measuring 110 degrees or higher in the summer of 2023.

Below: New development along the Valley Metro light rail line.

Phoenix has 51,020 acres of parks, according to the Trust for Public Land.

In 2023, Arizona officials announced limits on new groundwater-dependent development on the outskirts of the Phoenix metro area. But the state did not revoke approved permits, and Phoenix has plenty of projects in the works: in 2022, more than 550 buildings of five or more units were approved for construction in the metro area. Another 8,339 multi-family units were approved through the first half of 2023.

Different generations describe it differently. My dad tells me: if you can just do something about the heat in the summer here, you'll definitely be reelected. A different lens, but I think the same outcome.

AF: How has the pandemic affected your urban planning efforts? Did it surface any unexpected opportunities?

KG: The pandemic really changed how people interact with their communities. We saw biking and walking more than double. People tell us they hadn't realized how much they enjoyed that form of moving around their communities, and that they intend to keep some of those behavior changes. So we're looking at how we can create more public spaces. For instance, can we expand outdoor dining so that people can interact more?

Dr. Anthony Fauci has told us that the more time we can spend outdoors, the better for fighting COVID. But that has other great benefits. I serve as mayor of a city with more acres of parks than any other US city, and this has been a record year for us enjoying those Phoenix parks. You can be in the middle of Phoenix on a hiking trail and some days you don't see anyone else. So those amenities, and our focus on planning around parks, have really improved this year.

We also continue to invest in our transportation system. We've decided to speed up investment in transit—a decision that we had real debate over, and that I think will allow us to move toward a more urban form. We've actually seen increased demand for urban living in Phoenix. We have more cranes downtown than ever before and we're regularly seeing applications for taller buildings than we've ever seen here. I understand there's a national dialogue about whether everyone will want to be in a suburban setting, but the market is going in a different direction in our downtown.

COVID has also made us look at some of the key challenges facing our community, such as affordable housing, the digital divide, and addressing food insecurity, and we've made significant investments in those areas as well.

AF: Many might think of Phoenix as a place with abundant space for single-family homes, where a house with a small yard and driveway is relatively affordable. Yet the city has a big problem with homelessness. How did that happen?

KG: Phoenix competes for labor with cities like San Francisco, San Diego, and others that have much more expensive housing than we do, but affordable housing has been a real challenge for our community. Phoenix has been the fastest-growing city in the country. Although we have seen pretty significant wage growth, it has not kept up with the huge increases in mortgages and rent that our community has faced. It's good that people are so excited about our city and want to be part of it, but it's been very difficult for our housing market.

The council just passed a plan on affordable housing including a goal to create or preserve 50,000 units in the next decade. We are looking at a variety of policy tools, and multifamily housing will have to be a big part of the solution if we are going to get the number of units that we need. So again, that may be moving us toward a more urban form of development.

Above: Gallego (facing audience) greets the CEO of local nonprofit Native American Connections at the opening ceremony of a 63-unit affordable housing complex owned by the organization.

Phoenix's population grew 11.2% between 2010 and 2020, according to the US Census, making it the fastest-growing large city in the country over that decade.

As of early 2023, the city had created or preserved more than 29,000 housing units toward its goal of 50,000, including more than 10,000 affordable and work-force housing units.

AF: Opponents of the recent light rail expansion argued it would cost too much, but there also seemed to be some cultural backlash against urbanizing in that way. What was going on there?

KG: Our voters have opted time and time again to support our light rail system. There was a ballot proposition to ban light rail in 2019, shortly after I was elected. It failed in every single one of the council districts; it failed in the most Democratic precinct and the most Republican precinct in the city. Voters sent a strong message that they do want that more urban form of development and the opportunity that comes with the light rail system. We've seen significant investments in healthcare assets and affordable housing along the light rail. We've also seen school districts putting more money into classrooms and teachers' salaries because they don't have to pay for busing a significant number of students. We've really been pleased with its impact on our city.

When new businesses come to our community, they often ask for locations along the light rail because they know it's an amenity that their employees appreciate. So I consider it a success, but I know we're going to keep talking about how and where we want to grow in Phoenix.

AF: We can't talk about Phoenix and Arizona without talking about water. Where is the conversation currently in terms of innovation, technology, and conservation in the management of that resource?

KG: Speaking of our ambitious voters, they passed a plan for the City of Phoenix, setting a goal to be the most sustainable desert city. Water conservation has been a Phoenix value and will continue to be. The city already reuses nearly all wastewater on crops, wetlands, and energy production. We've set up strong programs in banking water, repurposing water, and efficiency and conservation practices, many of which have become models for other communities.

We are planning ahead. Many portions of our city are dependent on the Colorado River; that river system faces drought and may bring even larger challenges in the future. So we're trying to invest in infrastructure to address that, but also look at our forest ecosystem and other solutions to make sure that we can continue to deliver water and keep climate change front of mind. We've

In an agreement with federal and state agencies aimed at safeguarding the future of the Colorado River, the city will voluntarily forgo up to 50,000 acre-feet of its river allotment per year from 2023 through 2025, in exchange for up to $20 million or more per year to fund water conservation and resiliency measures.

also had good luck with using green and sustainable bonds, which the city recently issued. Partnerships with The Nature Conservancy and others have helped us look at how we manage water in a way that takes advantage of the natural ecosystem, whether through stormwater filtration or how we design our pavement solutions. So we've had some neat innovation. Many companies in this community are at the forefront of water use, as you would expect from a desert city, and I hope Phoenix will be a leader in helping other communities address water challenges.

Born and raised in Albuquerque, New Mexico, Kate Gallego received a degree in environmental studies from Harvard University and earned an MBA from the University of Pennsylvania. Before her 2013 election to the Phoenix City Council, Gallego worked on economic development for the Salt River Project, a not-for-profit water and energy utility that serves more than two million people in central Arizona. She has served as mayor of Phoenix since 2019.

Above: The 336-mile Central Arizona Project travels through Phoenix as it delivers water to nearly six million people in Arizona, more than 80% of the state's population.

The city implemented a Cool Pavement Pilot Program in 2020 in partnership with Arizona State University (ASU). By the end of 2023, the city expects to have treated 118 miles of pavement with a water-based asphalt treatment that reduces surface temperatures.

CAMBRIDGE

SUMBUL SIDDIQUI

In the Cambridge City Council chambers one week after being sworn in, 2020.

CAMBRIDGE

Sumbul Siddiqui immigrated to the United States from Karachi, Pakistan, at the age of two, along with her parents and twin brother. She was raised in affordable housing in Cambridge and educated in the city's public schools. Elected mayor of Cambridge in 2020, Siddiqui is the first Muslim mayor in Massachusetts. While in office, Siddiqui has advocated on behalf of the city's most vulnerable, striving to create affordable housing, protect households facing displacement, and promote equitable access to education. During the pandemic, she helped increase Internet access for low-income families and expanded free COVID testing for all Cambridge residents. Her agenda includes the promotion of clean and climate-resilient streets, parks, and infrastructure as part of making Cambridge a more equitable and civically engaged community.

Below: The Anderson Memorial Bridge over the Charles River. The river serves as the boundary between Cambridge and Boston.

ANTHONY FLINT: Cambridge has been gaining quite a lot of attention lately for a new policy that allows for some increases in height and density at appropriate locations—if the projects are 100 percent affordable. Can you tell us about that initiative and how it's playing out?

SUMBUL SIDDIQUI: The passing of the Affordable Housing Overlay was an important moment for me and for many on the city council. The proposal was to create a citywide zoning overlay to enable 100 percent affordable housing developments in order to better compete with market-rate development. The goal is to have multifamily and townhouse development in areas where they are not currently allowed. Our city has a widening gap between high- and low-income earners, and we always talk about diversity as a value. But how do we maintain that diversity? It's all about creating additional affordable housing options so more people can stay in the city. So now, many of our affordable housing developers, like our housing authority and other community development corporations, are holding community meetings about proposals; in some cases, they're able to add over 100 units to affordable housing developments that they were already going to build.

AF: Changes like this seem to percolate at the local level. I'm thinking, for example, of Minneapolis banning single-family-only zoning to allow more multifamily housing in more places, and several other cities followed suit. Is the 100 percent affordable overlay something that other cities might adopt, and did you anticipate that this might become a model for other cities?

SS: We certainly think that this can be a model. We know that our neighboring city, Somerville, is looking at it. It's all part of the overall mission, for many cities, to make sure that they're offering and creating more affordable housing options. This is housing that's affordable to your teachers, to your custodians, to your public servants, legal aid attorneys—you name it—so that they can stay in the city that they may have grown up in, or maybe they've moved out and want to come back. We want there to be that opportunity. We still see such stark inequality in our city; as someone who's grown up in affordable housing in Cambridge, I would not be here without it.

Cambridge's Affordable Housing Overlay allows developers to build permanently affordable housing at a greater height and density than is typically allowed under base zoning—and with faster approvals. Six projects that would add nearly 600 permanently affordable units were in the pipeline in 2023, with completion expected over the next one to three years.

AF: Cambridge has been such a boomtown for the last several years, and there has been a lot of higher-end housing development. Can you tell us about a few other policies that are effective in maintaining economic diversity?

SS: One way that we've been able to maintain our affordable housing stock is through the city's Inclusionary Housing Program. Under these provisions, developments of 10 and more units are required to allocate 20 percent of the residential floor area to low- and moderate-income tenants, or moderate- and middle-income home buyers. It's been an effective way to produce housing under these hot market conditions. The more people we bring to the city, the more we'll have that insatiable housing demand.

Below: Jefferson Park Apartments (left) and Frost Terrace (right) have won accolades for their approaches to affordable housing design.

In October 2021, the City of Cambridge announced an agreement with the owners of Fresh Pond Apartments, whose initial affordability terms were expiring, to preserve the affordability of 504 units for 50 more years.

We also want to focus on how we use city-owned public property that is available for disposition to develop housing. We've done a lot of work around homeownership options and making sure that we have a robust homeownership program for residents to apply to. Preservation is also a big part of the policy around affordability. This year, we've been working on the affordability of about 500 units in North Cambridge near the buildings I grew up in, and we've put in what will probably come to over $15 million to help preserve these market-rate buildings. Essentially, these are expiring use properties. So it's a little technical, but there are so many tools—and there's a long way to go.

AF: How did the pandemic reveal the disparities and racial justice issues that seem to be ingrained, in a way, in the economic outcomes of the city and the region?

SS: The pandemic has revealed a lot of the fault lines, and we've seen firsthand the disproportionate impact COVID has had on Black and Brown communities. It's highlighted longstanding issues around health care equity, and we've seen how so many of our low-income families have been unable to make ends meet. Many of them lost their jobs because of the public health crisis,

We always talk about diversity as a value. But how do we maintain that diversity? It's all about creating additional affordable housing options so more people can stay in the city.

but still needed to pay rent, pay utilities, and purchase food for themselves and their families. A lot of the issues we've seen during the pandemic have been issues all along, but as I've said, COVID has revealed those ugly truths even in our city . . . and you know, we can't turn a blind eye anymore.

And we've had to do things with more urgency. I always use the example of schools that had to close. We quickly got kids laptops and hotspots. Before the pandemic, we knew kids didn't have Internet at home, we knew kids didn't have computers, but we said, "Oh, you know, we're going to study that. . . ." We should have been doing these things all along. And so the pandemic has taught us to figure out solutions really quickly. It's shown us that we can make our city more accessible and affordable and we have to call out the injustices when we see them.

AF: The pandemic also arguably has been an opportunity to do some things with regard to sustainability, reconfiguring the public space. Could you talk about that and other ways you're helping to reduce carbon emissions and build resilience?

SS: This is an area where there's so much going on, yet sometimes it feels like we're not moving fast enough, given what we know. We are committed to accelerating the transition to net zero greenhouse gas emissions for all buildings in the city; our goal is net zero emissions by 2050. There are various types of incentives and regulations, and working groups are looking at questions like: how do we procure 100 percent of our municipal electricity from renewable sources? How do we streamline our efforts to expand access to energy-efficiency funding and technical assistance?

During the pandemic, Siddiqui started Rise Up Cambridge, one of the first guaranteed income programs in the country. The initial 18-month pilot concluded in March 2023, but the city has committed $22 million to extend the program for low-income families with children.

We're also revising our zoning ordinance to make sure that the sustainable design standards require higher levels of green building design and energy efficiency for new construction and major renovation. We're a city that loves our trees, right? So we are constantly looking at ways to preserve our tree canopy. We have a tree protection ordinance on the books that we are going to continue to strengthen this term. We continue to install high-visibility electric vehicle charging stations at publicly accessible locations. And there's a big push to incorporate green infrastructure into city parks and open spaces and street reconstruction projects. It's all hands on deck.

AF: The Lincoln Institute of Land Policy has called Cambridge home since 1974, when David C. Lincoln, son of our founder, chose to establish it in a place with world-renowned universities and other nonprofit organizations. Can you reflect on that distinctive feature of Cambridge—that is, the nonprofit, educational, medical, and other institutions being such a big part of the community?

SS: The universities in particular play a huge role. With the pandemic, I've seen our educational institutions, community organizations, small businesses, and residents work together to

address some of the most pressing issues. Thanks to the collaboration between the Broad Institute of Harvard and MIT and the City of Cambridge Public Health Department, we were the first in the state to offer COVID testing for residents and workers at all elder facilities. Now, we have seven-day-a-week testing in Cambridge. So that's a direct result of having the universities here in our space. And when we were setting up an emergency shelter for unhoused individuals, Harvard and MIT contributed funding toward that, and gave rent relief to retail and restaurant tenants that they have. So these partnerships have strengthened since the pandemic hit. They are such a big part of the community and they have risen to the occasion whenever I've called on them.

Above and facing page: A new crop of green buildings is growing in Cambridge. MIT.nano (left) is a 216,000-square-foot shared research facility widely hailed for its sustainable design. The King Open/Cambridge Street Upper Schools and Community Center (right) is the first net zero emissions school in Massachusetts.

Born in Pakistan and raised in Cambridge, Sumbul Siddiqui is a graduate of Brown University. She went on to serve as an AmeriCorps fellow at New Profit, a nonprofit organization dedicated to improving social mobility for families. After earning her JD from Northwestern University's Pritzker School of Law, Siddiqui returned to Massachusetts to work as an attorney with Northeast Legal Aid, serving the communities of Lawrence, Lynn, and Lowell. In 2020, she was elected mayor by the Cambridge City Council after serving as a councilor for two years; she was reelected in 2022, becoming the first female mayor in the city's history to serve two consecutive terms.

CLEVELAND

FRANK JACKSON

Cleveland native Frank G. Jackson, the city's longest-serving mayor (2006–2021), has long been an advocate for building equity and opportunity in this postindustrial city. During his tenure, Jackson focused on helping residents and businesses benefit from investments occurring in the city and advancing the Downtown Lakefront Development Plan. He also spearheaded Sustainable Cleveland 2019, a 10-year initiative designed to build a thriving regional economy, encourage new business practices, and improve air and water quality in this former manufacturing hub. Upon announcing in 2021 that he would not seek a fifth term, Jackson compared the job of being mayor to a relay race: "The race is not yet over and we are not yet a great city," he said, encouraging his audience to "ensure that the runner in the next leg of this race runs hard and he runs true."

Home to John D. Rockefeller's Standard Oil Company, Cleveland was once the center of the world's petroleum industry. The city's population more than quadrupled between 1850 and 1870, quadrupled again by 1900, and then doubled by 1920, at which point it was the fifth-largest city in the US.

ANTHONY FLINT: In the late 1800s, Cleveland was a booming place, arguably right up there with New York and Chicago—an incredible mix of innovation and jobs and homes and neighborhoods. Could you reflect on how that legacy has been on your mind as you've governed Cleveland over the last 15 years?

FRANK JACKSON: Well, it's always good to know history, so you can put yourself in the right frame of mind and have perspective. Cleveland was a booming place, with the Rockefellers and the economic successes of the Industrial Revolution. We were ideally located to be a hub, and for the distribution of goods and materials throughout the Midwest. So we reflect back on those heydays, fully recognizing that what brought us to that moment is no longer here. We need to look at where Cleveland is now and what could position our city to be in a similar situation as a hub for economic opportunity and prosperity and quality of life.

AF: At the statue in Public Square, former Mayor Tom Johnson is shown seated with his hand on a copy of *Progress and Poverty* by Henry George. Why do you think Cleveland was so receptive to the ideas of George, who believed the value of land should belong to everyone?

FJ: I couldn't tell you for sure, but as you know, the body takes its direction from its head . . . and I think Tom L. Johnson was a mayor with progressive thoughts and with the fortitude to execute and implement ideas. So he wasn't just a conversationalist, he actually did things.

Cleveland was in transition then—fast forward, and we're in the same kind of transitional period. The Industrial Revolution produced a certain level of prosperity and wealth, but also a certain social condition that I believe that progressive era was attempting to change, in order to create more equitable outcomes.

I admit, I haven't really studied Mr. George's philosophy. But what I do understand is this progressive notion of land use, and how land should not be controlled by a few entities. There should be broader input into what happens on that land.

AF: As the city has steadily emerged from a period of decline and population loss during the second half of the 20th century, what have been the critical elements of its regeneration? What catalysts are you most hopeful about?

Above: A statue of former Cleveland Mayor Tom L. Johnson in the city's Public Square shows him holding a copy of Henry George's *Progress and Poverty*. Published in 1879, the book questioned why poverty existed alongside the great wealth of the era. Wildly popular in its day and one of the best-selling political economics books of all time, it inspired the founding of the Lincoln Institute of Land Policy.

Facing page: Industrial buildings
along the Cuyahoga River in
the early 1900s (top). The city
has changed dramatically, but
its industrial character (bottom)
is still a defining feature.

FJ: Well, it's how you position yourself—how does Cleveland position itself for the future? How do we ensure a sustainable economy? How do we deliver goods and services? And how do we get into sustainable industries, like electric vehicles. . . . All of this includes technology, all of it includes education, all of it includes research and development. All these things are inclusive of each other. So there's not just one thing we can pick and say we're going to do.

I think we need to go back to what Henry George was talking about, and what Tom L. Johnson was trying to do, which is to say that progress is only sustainable if we have equity, and if we eliminate the disparities and inequities in the way our social, political, and economic systems function. And as you know, particularly around the social unrest these days, if we fail to address issues of classism and racism, then all our efforts will be doomed.

AF: Race and economic development are very much on every mayor's mind these days, especially now that the pandemic has revealed so much entrenched inequity. What are some of the most effective ways Cleveland has addressed historical segregation and racial disparities?

FJ: Before I answer that, let me just say that whatever we have done is not sufficient. . . . So we've taken the step of declaring violence and poverty as public health issues, and establishing a new division in the Department of Health around social justice. We're trying to institutionalize some things.

We have also attempted to work with our private sector partners to address inequities, disparity, and racism within their organizations. We're helping them to have a better outcome in terms of contracting for goods and services with lending institutions; even though redlining is illegal, the actual practice of how investments are made and moneys are lent and developments occur is basically redlining. We try to work with them to help them take risks where they normally would not take risks. That can only happen if you allow for wealth to occur among those who have traditionally been denied wealth; if you have leadership and career opportunities for those who have traditionally been denied those opportunities. So those are the kinds of things that we work on.

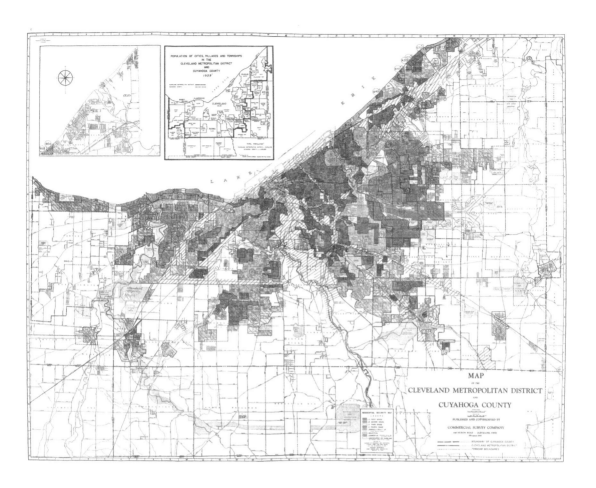

Above: Federal redlining maps from the 1930s rated the "hazardous" nature of neighborhoods in Cleveland and other cities, based largely on the race of residents. Those ratings correlate to health and economic outcomes today.

Jackson oversaw the first revisions to Cleveland's zoning code since 1929, including streamlining downtown height requirements, and planted the seeds for significant shifts such as piloting a simpler form-based code to encourage mixed-use development and eliminating minimum parking requirements.

The real question is, what is the culture of Cleveland? How does Cleveland function, and what is its attitude toward these issues? And that's a behavioral thing that bureaucracy cannot really regulate.

AF: Can you tell us about recent zoning reform measures aimed at reducing barriers to housing production and other local economic activity? How important are these rules and regulations to regeneration, and how has Cleveland made innovative use of vacant and abandoned land?

FJ: Land use is key. We're moving toward having zoning more aligned with people and multiple mobility, where we have bikes, cars, scooters, walking, jogging. In trying to create that type of

Whatever we do, it will never be sustainable if we don't address the underlying issues that are the issues of America: institutionalized inequity, disparities, racism, and classism, which have a lot to do with land.

city, it's important to have zoning that will accommodate it in a way that minimizes conflict.

When I first came into government, there was no new housing development in Cleveland. As a result of the negative impacts of federal and state policy around redlining and urban renewal and then the social impact of riots, we had acres and acres of vacant land in the central city, predominantly in African American communities. Mayor Michael White, who led the city from 1990 to 2001, was a genius in this regard. He worked with the financial institutions and developers to create a network of neighborhood nonprofits whose primary purpose was to redevelop land for housing and to redevelop land at all price ranges, that would make it affordable. I'm familiar with it because I was councilman of Central, where I still live, which probably had the most negative impacts.

We continue this effort today with Rescue Plan Act money; we're getting $511 million and we're working with the private sector to develop tools. We're not talking about a project or initiative; we're developing tools. We're working to connect all these dots—a lot of that has to do with land and with the availability of land, whether it's lakefront land or empty office space downtown or warehouses, old industrial sites that need environmental cleanup. It's not just housing, but also, how do we create entrepreneurship, commercial strips, retail strips that still have the bones—how do we bring them back and have ownership of goods and services being provided to the community by the people in that community or someone who looks like the people of that community?

Like other Black communities across the country, Central suffered from decades of disinvestment after receiving a hazardous rating on federal redlining maps in the 1930s. After being elected to the City Council in 1989, Jackson led a redevelopment effort there, securing federal Homeownership Zone funding that made it possible for the city to leverage other financial support, build hundreds of new houses on vacant parcels, and rebuild public housing.

Cleveland received the eighth-largest allocation of American Rescue Plan Act (ARPA) funds in the nation. The first installment arrived several months before the end of Jackson's final term, and his administration worked with the Cleveland City Council to allocate funding for projects including neighborhood redevelopment and support for small businesses, homebuyers, and renters.

AF: Well, if there's one thing that Cleveland has, it's good bones, right?

FJ: That's exactly right. One of the things that culturally came out of that period that you talked about, the heyday of Cleveland, was Severance Hall—home of the Cleveland Orchestra—the museums, the whole University Circle area. Now we're trying to use old industrial sites and lakefront or riverfront property in new ways since they're no longer used for commerce . . . but the freeway, railroad tracks, those kinds of things are almost impossible to remove. They're barriers. So how do you overcome those barriers? One of the things we're looking at is a land bridge that would allow for green space and access to the riverfront, the lakefront, and with that to always have public access and not have private ownership of the waterfront.

AF: Sounds like there's a lot of reimagining going on.

FJ: That's the advantage to where Cleveland is now. To have a blank canvas, so to speak, gives us that opportunity. Now the question is whether or not we mess it up. . . . Whatever we do, it will never be sustainable if we don't address the underlying issues that are the issues of America: institutionalized inequity, disparities, racism, and classism, which have a lot to do with land.

Above: The Lake Erie waterfront, whose attractions include (left to right) the Cleveland Browns stadium, Great Lakes Science Center, and Rock & Roll Hall of Fame, has been cut off from Cleveland's downtown by roads and railroad tracks since 1849. Local civic and municipal leaders have long advocated for a land bridge that would improve pedestrian access by reconnecting city and coast.

Facing page: Participants in the 30th annual Chalk Festival at the Cleveland Museum of Art in 2019.

Frank G. Jackson is a lifelong resident of Cleveland's Central neighborhood, where he began his career in elected office as a City Council member before becoming City Council president. A graduate of Cleveland Public Schools, Cuyahoga Community College, and Cleveland State University—from which he earned bachelor's, master's, and law degrees—Jackson began his public service career as an assistant city prosecutor in the Cleveland Municipal Court Clerk's Office. Elected as the 56th mayor of Cleveland in 2005, Jackson was reelected three times before announcing he would not seek a fifth term in 2021.

BOGOTÁ

CLAUDIA LÓ

Field visit to the Sendero Chamicero, an ecological trail that will connect the hills of Bogotá to the Torca-Guaymaral Wetlands, 2022.

Claudia López was elected mayor of Bogotá in October 2019, after campaigning with a focus on climate change and other environmental and social issues. She is the city's first elected female mayor and first openly gay mayor. Before entering politics, she worked as a researcher and helped expose connections between high-level politicians and paramilitary groups in Colombia, a national scandal that came to be known as "parapolitics." López is on the steering committee of C40, a global network of mayors united for climate action, and belongs to UN-Habitat's Council on Urban Initiatives, which advocates for healthy, just cities. She gave this interview on her way to the COP26 climate summit in Glasgow.

Facing page: Downtown Bogotá from Monserrate. The cable-car system has made this view possible since 1955.

López and the Bogotá City Council declared a climate emergency in December 2020, becoming the first Latin American city to do so. "This structural emergency requires urgent mitigation and adaptation actions, considering climate change as the focus of all decisions, strategies, and planning tools," reads the first article of the 14-page declaration.

ANTHONY FLINT: Your victory suggests that residents are ready for serious action on the environment and climate change. Do you feel you have a mandate, and what are your top priorities in terms of climate?

CLAUDIA LÓPEZ: Well, there is no doubt that I have a clear mandate from Bogotá's people. During my campaign, I made a public commitment to environment and climate change issues. We have a deep social debt and a deep environmental debt that we have to pay. After the pandemic, the social debt will be harder to address than the environmental debt, because COVID has doubled unemployment and poverty in my city. On the other hand, I am still very optimistic that postpandemic opportunities to address environmental issues will increase.

We have to adapt—that's our mandate. In the context of Colombia, we have three general issues. One of them, and the major contributor to climate change, is deforestation. This is an issue mainly for rural Colombia, and is by far our country's largest contribution to

the environmental crisis and the climate emergency. The second factor is fossil fuels; transportation is Colombia's second-largest contribution to the climate emergency. And the third is related to waste management.

What are we doing? Migrating from a monodependent diesel bus system toward a multimodal system based on a metro, a regional train system, cable system, and buses. We're also transforming our approach to waste management into a recycling, green, circular economy, so that we can turn waste into clean energy. We're making the city greener. Building cities has always been about hardening rural and green areas, but what we need to do in the 21st century, I think, is the opposite—we need to take advantage of every public space that we have, making every effort not only to plant trees, not only to plant gardens, but to transform urban areas, gray areas, into green areas.

CALLE

What makes a city safer? The first thing I think is to make the city sustainable, and that means greener, and that means more equitable.

We're lucky that we have the legal mandate to propose a new master plan, the POT—*Plan de Ordenamiento Territorial*. We can include these changes and investments, not in a four-year-term government plan, but in a 14-year city plan. We are trying to take advantage of this moment.

AF: This year marks the centenary of the Colombian value capture tool *contribución de valorización*, or betterment contributions. What is your vision for building on that tradition?

CL: I think that's critical. The most important financial tool we have for sustainable development is land value capture. In our POT, we are including not only the traditional betterment contribution, but also many other ways to use land value capture.

We have at least seven different tools, financial tools, all related. Basically, we determine the value that's going to be generated by a transformation of land use and we agree with the developer, so that the developers don't pay us in cash, as in the betterment contribution, but pay by building the infrastructure and the urban and social equipment that new development will need.

This is not about having lovely maps with marvelous plans; it's about having the money to redistribute the cost and benefit of sharing and receiving. That is actually what I think urban planning is: making sure that either through public investments or through land value capture or through private investments, we ensure an equitable and sustainable share of the cost and benefits of building the city. That's the role of the government, and that's what we're trying to achieve here.

Bogotá's master plan focuses on four pillars: sustainable mobility, including active transport; urban greening; a recognition and redistribution of the time devoted, especially by women, to unpaid care work; and a sustainable economic recovery.

Betterment contributions are fees paid to municipalities by owners of select properties to help defray the costs of public improvements or services from which they directly benefit. They are a form of land value capture, a policy approach that enables communities to recover and reinvest increases in property values.

AF: I'd like to turn to the topic of crime. How has the problem of crime had an effect on the perception of the city, and particularly on the perception of public space in the city?

CL: It has had a huge impact, of course. The more crime you have in public spaces, as a fact or as a perception, the less well-being you have as a city. What makes a city safer? The first thing I think is to make the city sustainable, and that means greener, and that means more equitable.

My top priority for making Bogotá safer is not to add cameras and technologies; it is to make sure that the city has the capacity to provide fair and legal employment for our population, especially our youth. The social roots of security are more important.

I'm very excited and very proud that we're building into our POT and our land use development plan criteria for women and caregivers as criteria for urban development. If you can make a city safer for women, if you can make a city safer for kids, that will be a city safer for everyone.

Now the second thing, as important as transportation, infrastructure, and social infrastructure in the 21st century, is digital infrastructure. We are going to extend *fibra óptica*, the best, fastest Internet, to every neighborhood in our city, to every school in our city. That's crucial to make a more sustainable, more equitable, and safer city. At this postpandemic moment, we're having a severe backlash in insecurity in our cities. It's not only in Bogotá, it's global. Unfortunately, higher unemployment and higher poverty always correlate with higher insecurity.

AF: Which policies are working to make life better in informal settlements, such as upgrading or infrastructure, and what in your view needs to change?

CL: We have at least three innovations included in our land use plan that I'm very proud of. As you know, in Latin America, roughly half of our cities have been built informally. Our land use development plan is the first that clearly assumes that, and accepts that. Instead of making a land use plan that is only useful for the formal city, for half of the city, we have a plan that recognizes that 45 percent of our city is informal. It creates an urban norm,

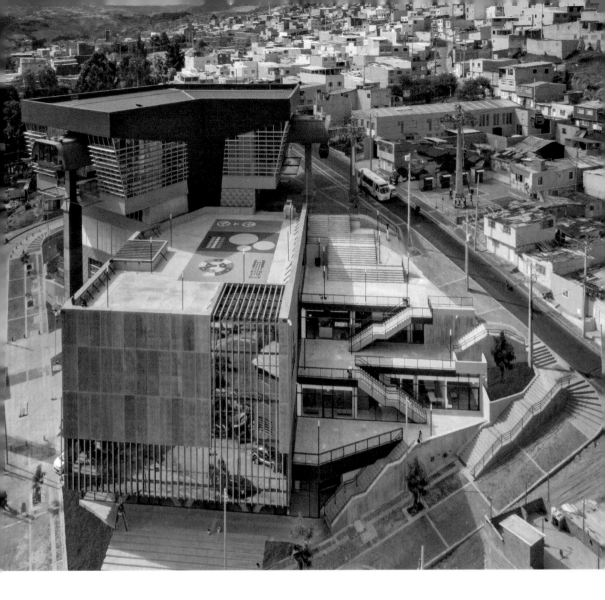

urban rules, and an urban institution to help people improve their homes in the informal city, and to improve their neighborhoods. It includes all people within the land use development plan.

In Bogotá, we have an institution called *curaduría*, which provides urban licensing and construction licenses. We are creating a public *curaduría* for the informal city. There's no way that you can impose an urban standard on half of the city that they don't have any chance to meet. We also have the *Plan Terrazas*, which says, after we improve your first floor, after we improve it properly, then you can build your second floor, for example, or you can

As part of the *Curaduría Cero* program, started by the Colombian government in 2017, López launched the *Plan Terrazas* in Bogotá in 2020. The goal is to formalize and fortify the city's informal, earthquake-vulnerable houses, improving safety and comfort.

build some space for economic activity in your first floor. You will improve your housing, but you will also improve your income. For poor people, housing is not only a place to live, it's also a place to produce and generate income.

The second important thing we created is this Care system, thinking particularly about women. Half of the economy is informal. That means people don't have formal jobs with pension funds and health insurance; they don't have care when they are sick or aging. So who takes care of the sick and elderly? It's the unpaid women who do that; roughly 1.2 million women in Bogotá don't have jobs, don't have education, don't have time for themselves because they are caregivers. For the first time in Bogotá, we are reserving land for social infrastructure to provide institutional health care. For children, for women, for elders, for people with disabilities, so that we can relieve women, so they can access time to rest. They don't have a free week ever in their lives.

Facing page: The *Plan Terrazas* helps families improve their dwellings (top); nearly half of Bogotá is made up of informal settlements (bottom).

We have to adapt—that's our mandate.

We're trying to balance. I think the development in Bogotá has been incredibly unbalanced, with much of the advantage on the developers' side. Of course, the developers need profitability, and we are trying to find the equilibrium point.

Claudia López was a senator of the Republic of Colombia from 2014 to 2018, becoming a prominent figure in the fight against corruption; she was the vice presidential candidate for the Green Alliance party in the 2018 presidential election. In 2019, she became the first popularly elected female mayor of Bogotá. Prior to politics, López worked as a journalist, researcher, and political analyst. She studied finance, public administration, and political science at the Universidad Externado de Colombia, and went on to earn a master's degree in public administration and urban policy from Columbia University and a PhD in political science from Northwestern University.

BIRMINGHAM

RANDALL W

Connecting with a resident in the Gate City community, 2020.

OODFIN

BIRMINGHAM

When he was elected in 2017 at age 36, Randall L. Woodfin became the youngest mayor to take office in Birmingham, Alabama, in 120 years. Reelected in 2021, Woodfin has made revitalization of the city's 99 neighborhoods his top priority, along with fostering a climate of economic opportunity and leveraging public-private partnerships. In a city battered by population and manufacturing losses, including of the iron and steel industries that once thrived there, Woodfin has looked to education and youth as the keys to a better future. He established Birmingham Promise, a public-private partnership that provides apprenticeships and tuition assistance to cover college costs for high school graduates, and launched Pardons for Progress, which removed a barrier to employment opportunities through the mayoral pardon of 15,000 misdemeanor marijuana possession charges dating to 1990.

Below: Founded at the intersection of two railroads in 1871, Birmingham grew quickly, earning the nickname the Magic City.

ANTHONY FLINT: How do you think your vision for urban revitalization played into the large number of first-time voters who've turned out for you?

RANDALL WOODFIN: My vision for urban revitalization—which, on the ground, I call neighborhood revitalization—played a significant role in not just the usual voters coming out to the polls to support me, but new voters as well. I think they chose me because I listen to them more than I talk. Many residents have felt, "Listen, I've had these problems next to my home, to the right or to the left of me, for years, and they've been ignored. My calls have gone unanswered. Services have not been rendered. I want a change." I made neighborhood revitalization a priority because that's the priority of the citizens I wanted to serve.

AF: With the Infrastructure Investment and Jobs Act and the American Rescue Plan Act bringing unparalleled amounts of funding to state and local governments, what are your plans to distribute that money efficiently and get the greatest leverage?

RW: This is a once-in-a-lifetime opportunity to really supercharge infrastructure upgrades and investments we need to make in our city and community. This type of money probably hasn't been on the ground since the New Deal. When you think about that, there's an opportunity for the city of Birmingham citizens and communities to win.

We set up a unified command system to receive these funds. In one hand, in my left hand, the city of Birmingham is an entitlement city and we'll receive direct funds. In my right hand, we have to be aggressive and go after competitive grants for shovel-ready projects.

With our Stimulus Command Center, we partner not only with our city council, but also with our transportation agency. We have an inland port, so we partner with Birmingham Port. We also partner with our airport and our water works department. All of these agencies are public agencies that happen to serve the same citizens I'm responsible for serving. For us, a collective approach to all these infrastructure resources is the best way. We have an opportunity with this funding not only to supercharge our economic identity, but to make real investments in the infrastructure that our citizens use every day.

Woodfin's 2023–2024 budget proposal called for increases in neighborhood revitalization efforts such as street resurfacing, sidewalk repair, and weed abatement, and the city was able to fund capital improvements to libraries, parks, and recreation centers thanks to a budget surplus of $81 million remaining from fiscal year 2022.

The $1.9 trillion American Rescue Plan Act (ARPA) provided more than $65 billion in flexible funding to cities and towns, including more than $140 million to Birmingham, to spur economic recovery from the COVID pandemic. The $1.2 trillion Infrastructure Investment and Jobs Act, meanwhile, will provide states and cities with approximately $550 billion in new infrastructure funding over 10 years.

Above: In July 2023, HUD Secretary Marcia Fudge, right, came to Birmingham to announce that the city would receive a $50 million neighborhood revitalization grant. Woodfin joined Fudge and US Rep. Karen Sewell, left, for the event.

Facing page: The Fourth Avenue District, one of the only surviving Black business corridors in the Southeast, was selected to join a state revitalization program in 2019.

AF: What in your view are the key elements of neighborhood revitalization and community investment that truly pay off for legacy cities?

RW: This is how I explain everything that happens from a neighborhood revitalization standpoint: I first share the problem through a story. The city of Birmingham is fortunate to be made up of 23 communities in 99 neighborhoods. Consider going to a particular neighborhood in a particular block. You have a mother in a single-family household where she is the responsible breadwinner and owner. She has a child or grandchild that stays with her. She walks out onto her front porch. She looks to her right, and there's an abandoned, dilapidated house that's been there for years that needs to be torn down. She looks to her left, and there's an empty lot next to her. When she walks out to that sidewalk, she's afraid for her child or her grandchild to play or ride a bicycle on that sidewalk because it's not bikeable. That street, when she pulls out from the driveway, hasn't been paved in years. The neighborhood park she wants to walk her child or grandchild down to hasn't had upgraded, adequate playground equipment in some time. She's ready to walk her child or grandchild home because it's getting dark, but the streetlights don't work. Then she's ready to feed her child or grandchild, but they live in a food desert. These are the things we are attempting to solve.

One is blight removal, getting rid of that dilapidated structure to the right of her. We need to go vertical with more single-family homes that are affordable and market rate so we don't have "snaggletooth" neighborhoods where you remove blight, but then have a house, empty lot, house, empty lot, empty lot.

That child—we have to invest in that sidewalk so they can play safely or just take a walk. We have to pave more streets. We have to have adequate playground equipment. We have to partner with our power company to get more LED lights in that neighborhood, so people feel safe. We have to invest in healthy food options so our citizens can have a better quality of life. These are the things related to neighborhood revitalization that I frame and address to make sure people want to live in these places.

AF: What are your top priorities in addressing climate change? How does Birmingham feel the impacts of global warming, and what can be done about it?

RW: Climate change is real. We're not near the coast and so we don't feel the impact right away that other cities do, like Mobile would in the state of Alabama. However, when certain weather events happen on the coast in Alabama, they do have an impact on the city of Birmingham. We also have the issue of tornadoes, which are increasing over the years; they affect a city like Birmingham that sits in a bowl in the valley.

Birmingham spent $8 million repaving city streets in 2022 and invested another $12 million in repaving 45 miles of road segments across the city in 2023.

I made neighborhood revitalization a priority because that's the priority of the citizens I wanted to serve.

Birmingham was founded from a blue-collar standpoint of iron and steel and other things made here. Although that's not driving the economy anymore, there are vestiges of those industries that have a negative impact. We have a Superfund site right in the heart of our city that has affected people's air quality, which is totally unacceptable. So addressing climate change from a social justice standpoint has been a priority for the city of Birmingham and this administration. We are partnering with the EPA for our on-the-ground local issues.

From a national standpoint, Birmingham has joined other cities as it relates to the Paris Agreement, but this conversation on climate

Facing page: The Sloss Furnaces
complex (top), which was originally
constructed in 1881 and became
the world's largest manufacturer
of crude iron, is a National Historic
Landmark that pays tribute to the
city's industrial past. Sloss and
other companies left another kind
of legacy: the 35th Avenue Superfund site, shown in 1972 (bottom),
encompasses three neighborhoods
in North Birmingham heavily pollut-
ed by coal, asphalt, and steel plants.
The Environmental Protection
Agency has remediated 650 properties to date, removing 90,000 tons
of contaminated soil.

Land banks can acquire abandoned
or tax-delinquent properties and
prepare them for redevelopment or
community ownership by clearing
their titles of any back taxes, municipal liens, or other outstanding
liabilities. The Birmingham Land
Bank Authority, created in 2014,
has cleared more than 530 titles to
date, and recently launched an
Accelerated Home Ownership pilot.

change can't be in the isolation of a city. Unfortunately, the city
of Birmingham doesn't have home rule, and having the conversations with our governor—about the importance of the state of
Alabama actually championing this issue and joining calls of, "We
need to make more noise and be more intentional and aggressive
about climate change"—has been a struggle.

AF: What about your efforts to create safe, affordable housing,
including a land bank?

RW: I look at it as a toolbox. Within this toolbox, you have various
tools to address housing. At the height of Birmingham's population, in the late sixties, early seventies, there were about 340,000
residents. Now we're down to 206,000 residents within our city
limits. You can imagine the cost and burden that's had on our
housing stock. And when you add to that the homes that pass
from one generation to the next and are not necessarily being
taken care of, we've had a considerable amount of blight.

Like other cities across the nation, Birmingham has a land bank.
This land bank was created prior to my administration, but we've
attempted to make it more efficient. We're driving that efficiency
not just by looking toward those who can buy land in bulk, but
also by empowering the next-door neighbor, or the neighborhood,
or the church that's on the ground within that neighborhood, to
be able to participate in purchasing the lot next door. This helps
to make sure, again, that we can get rid of these snaggletooth
blocks or snaggletooth neighborhoods and go vertical with
single-family homes.

We're also acknowledging that in urban cores, it's hard to get
private developers to the table. With some of our ARPA funds,
we've been setting aside money to offset some of these developer costs to support not only affordable, but market-rate housing
within our city limits. We want to make sure our citizens have
a seat at the table, and that they feel empowered—that there's
a path for them if they want to have a home.

AF: Finally, tell us a little bit about your belief in guaranteed
income, which has been offered to single mothers in a pilot
program. You've joined several other mayors in this effort.
How does that reflect your approach to governing this midsize
postindustrial city?

With the help of a $500,000 grant from Mayors for a Guaranteed Income, Woodfin piloted a 12-month basic income program called Embrace Mothers. From March 2022 to February 2023, 110 single mothers received $375 per month on a prepaid debit card. Participants spent the largest share of the money (32%) on food and groceries. More than 48 US cities started guaranteed income programs between 2020 and 2022, according to the *New York Times*.

In 2023, according to the US Census Bureau, about 23% of children under the age of 18 in the US live in single-parent households. Of these 10 million single-parent families, almost 80% are headed by single women, a third of whom live in poverty.

RW: The city of Birmingham is fortunate to be a part of a pilot program that offers guaranteed income for single-mother families in our city. This income is $375 a month over a 12-month period—no strings attached, no requirements on how they can spend the money.

Every city in this nation has its own story, its own character, its own set of unique challenges. At the same time, we all share similar fates and have similar issues. The city of Birmingham has its fair share of poverty. We don't just have poverty, we have concentrated poverty, and guaranteed income is another tool within that toolbox for reducing poverty. Over 60 percent of Birmingham's households are led by single women. That is not something I'm bragging about. That is a fundamental fact. A lot of these single mothers struggle.

I think we all would agree, no one can live off $375 a month. If you had this $375 additional funding in your pocket or your homes, would that help your household? Does that help keep food on

the table? Does it help keep your utilities paid? Does it help keep clothing on your children's backs and shoes on their feet? Does it help you get from point A to B to keep your job to provide for your child?

This is why I believe this guaranteed income pilot program will be helpful. We only have 110 slots, so it's not necessarily the largest amount of people, but I can tell you over 7,000 households applied. The need is there for us to do every single thing we can to provide more opportunities for our families to be able to take care of their families.

Above: Railroad Park, a 19-acre green space known as "Birmingham's living room." Facing page: The Alabama Theatre, a downtown institution since 1927.

Born and raised in Birmingham, Randall Woodfin left the city to earn a degree from Morehouse College, then came back to work for the Birmingham mayor, the city council, and the Jefferson County Committee on Economic Opportunity. After earning a law degree from Samford University's Cumberland School of Law, he worked as an assistant city attorney and political organizer, and was elected president of the Birmingham Board of Education in 2013. Woodfin was elected mayor in 2017 and reelected in 2021.

MIRO WEIN

BERGER

Groundbreaking at the Moran Frame, a public art project created from an abandoned coal plant in Water Works Park, 2020.

A native Vermonter who was first elected in 2012, Miro Weinberger is serving his fourth term as the mayor of Burlington. Vermont has long been a progressive kind of place, with a population dedicated to environmental measures, whether solar and wind power, electric vehicles, or sustainable farming practices. Burlington, its change-agent hub—the place that gave rise to Bernie Sanders, who served as mayor from 1981 to 1989—became the first city in the country to source 100 percent of its energy from renewables in 2014. Since then, Weinberger and other leaders have continued to build on that foundation, taking more and broader steps to shift the city's energy, transportation, and building sectors away from fossil fuels entirely, with the goal of becoming a net zero energy city by 2030.

Right: Pedestrians explore Church Street, Burlington's central outdoor marketplace, in 2020.

ANTHONY FLINT: Tell us about this ambitious goal of becoming a net zero energy city by 2030. What is that going to look like, and what are the steps to make that happen?

MIRO WEINBERGER: As a result of decades of commitment to more efficient buildings and weatherization, Burlington uses less electricity as a community in 2022 than we did in 1989, despite the proliferation of new electrical devices and what-not. That sounds exceptional, and it is. If the rest of the country had followed that trajectory, we'd have something like 200 less coal-burning plants today.

When we became a 100 percent renewable electricity city in 2014, there was enormous interest in how Burlington had gotten there. After talking to film crews from South Korea and France and answering question after question about how we did this, I came to think we had achieved it for two big reasons. First, there was political will. Second, we had a city-owned electric department with a lot of technical expertise that was able to make this transformation to renewables affordable.

The way we are defining net zero is to use no fossil fuels—or to have a net zero fossil fuel use—in three sectors. For the electricity sector, we're already there. That gets us about 25 percent toward the total goal. The other two are the ground transportation sector and the thermal sector—how we heat and cool our buildings.

The big strategies are electrifying everything: electrifying all the cars and trucks that are based here in Burlington, and moving the heating and cooling of our buildings to various electric technologies, the most common one probably being cold-climate heat pumps. Then, rounding out the strategies, we are looking to implement a district energy system that would capture waste heat and use it to heat some of our major institutional buildings. We're also making changes to our transportation network to make active transportation account for more of our vehicle trips and bring down fossil fuel use that way as well. Those are the major roadmap strategies.

AF: Is there one component of this goal that you have found particularly tough in terms of trying to go citywide?

In 2023, Burlington applied for state approval for an energy system that would pipe excess steam from the wood-fired McNeil Generating Station power plant to heat the nearby University of Vermont, the UVM Medical Center, and the non-profit Intervale Center. As designed, the project would cut the city's natural gas emissions by 16%.

The city has invested in active transportation networks with bike lanes, sidewalk reconstruction, and other projects to improve safety and access for bicyclists and pedestrians. A 2023 update on the city's progress toward net zero reported a reduction in vehicle miles traveled.

After ramping up rebates for electric vehicle purchases and home efficiency upgrades, Burlington saw a fivefold increase in residential heat pump installations, projected to reduce the city's annual carbon emissions by 4.5 million pounds per year. The city introduced new and expanded incentives in 2023.

MW: In general, I've been really pleased with our progress. We actually found, in our first update in 2021, that we were on target to meet this incredibly ambitious goal of phasing out fossil fuels by 2030. Part of that, admittedly, was that 2020 was a pretty exceptional year, as we all know, and we did see transportation-related emissions drop as a result of the pandemic. We just got a new measurement and we did see some rebounding, so we are not quite on track after two years the way we were after one. But the rebound here in Burlington was only about a quarter of the nationwide rebound in emissions. Basically, we had a 1.5 percent increase in emissions after the pandemic, whereas the rest of the country grew by 6 percent. We've seen a rapid increase in the adoption of heat pumps and electric vehicles over the last couple of years since we came forward with what we call green stimulus incentives very early in the pandemic.

That said, I often have this sensation that we are fighting this battle with one hand tied behind our back, because it is not a level playing field for new electrification and renewable technologies.

This chart represents energy generated and purchased by the Burlington Electric Department in 2022 prior to renewable energy certificate sales, totaling 353,459 MWh of electricity (all figures are rounded). Burlington Electric has no contracts for resources fueled by natural gas, nuclear, or coal.

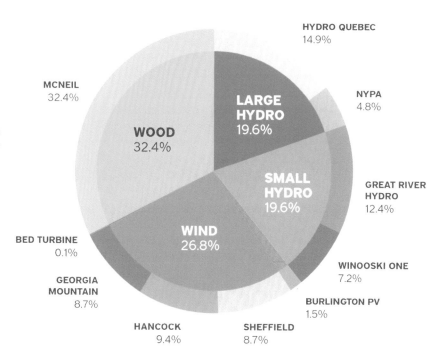

HYDRO QUEBEC
14.9%

MCNEIL
32.4%

LARGE HYDRO
19.6%

NYPA
4.8%

WOOD
32.4%

SMALL HYDRO
19.6%

GREAT RIVER HYDRO
12.4%

WIND
26.8%

BED TURBINE
0.1%

WINOOSKI ONE
7.2%

GEORGIA MOUNTAIN
8.7%

BURLINGTON PV
1.5%

HANCOCK
9.4%

SHEFFIELD
8.7%

The costs of burning fossil fuels are not properly reflected in the economics right now. We need a price on carbon in some form. The fact that we don't have that holds us back. When we get that—and I do think it's inevitable that eventually we will get this policy right, like a growing number of jurisdictions around the world—I think we're going to have a wind at the back of all these initiatives. It will help with everything we're trying to do.

AF: Now, I want to make sure I understand. Do you want everyone in the city of Burlington to operate an electric vehicle by 2030? Is it that kind of scaling up and adoption?

MW: Basically, yes. That is what it would really take to fully achieve the goal—that or some offset investments to help us get there. We are very serious about doing everything we can to bring about this transformation as quickly as possible.

Every time we bring another electric vehicle or heat pump online, that's a new revenue stream to the city.

A year ago, we passed a zoning ordinance that says new construction in Burlington cannot burn fossil fuels as the primary heating source. We didn't prohibit fossil fuels—we thought that was too onerous, and the technology's just not there to go that far. Regulating the primary heating source can bring down the impact of a new building by as much as 85 percent. In recent weeks, the state signed off on a change to our charter that gives us the ability to go beyond that and put new regulations in place for all buildings in Burlington. By next Town Meeting Day, we plan to have a new ordinance in front of the voters that would start to put requirements in place for the transformation of mechanical systems for major new and existing buildings when they get to the end of their useful life. When water heaters break, for example, we are going to have this strategy through our utility offering very generous incentives, and have actual regulatory standards in place that require transformation.

Burlington's Carbon Pollution Impact Fee and new thermal heating requirements passed in March 2023. Beginning in 2024, new construction must be fully renewable, and city buildings and large existing buildings must use renewable heating and water heating systems whenever older systems are replaced.

AF: I want to ask about the utilities. You mentioned Burlington Electric, the city-owned utility, and then, of course, you have Green Mountain Power. How important is their role, given that utility companies elsewhere seem to be wary of renewables and may even end up hindering that transition?

Above: Buying the 7.4-megawatt hydro station Winooski One in 2012 helped Burlington reach its goal of 100% renewable energy.

Facing page: The lights of downtown Burlington extending to the shore of Lake Champlain.

In 2022, *Fast Company* named Green Mountain Power one of the 10 most innovative companies in North America for "flipping the switch on energy generation," citing its pioneering work on microgrids, battery storage, and resiliency zones.

MW: I've got to say, a decade in office grappling with these issues has made me a big believer in publicly owned power. All of the work that I described over the last 30-plus years—the city-owned electric department has been a big part of that. Municipalities, towns, mayors that don't have their own electric utility—I think it's harder. There are things that any local community can do to collaborate with and, when necessary, bring public pressure to bear on utilities, which tend to have to answer to some public regulatory authority. I think that there are ways to push other utilities to do what Burlington Electric is doing. It's an exciting story in Vermont that the other utility that has been quite innovative, Green Mountain Power, is an investor-owned utility.

If we get anywhere near this net zero goal, it's going to mean we're selling a whole lot more electricity than we are now; we estimate at least 60 percent more than today. Every time someone buys an electric vehicle and charges it up in Burlington now, and they do it at night, we're able to sell them off-peak power. That brings more dollars into the utility. It's very good, the economics; that's why we're able to offer these generous incentives. Every time we bring another electric vehicle or heat pump online, that's a new revenue stream to the city. These incentives largely pay for themselves with that new revenue. To me, it seems like good business sense to move in this direction.

AF: Vermont has become a very popular destination for mostly affluent climate refugees who are buying up land and building houses. What are the pros and cons of this trend?

Vermont gained nearly 5,000 net new residents from July 2020 to July 2021. In addition to offering a safe escape for city dwellers fleeing the pandemic, the state is attracting transplants based on its reputation for having a relatively low climate risk: the Environmental Protection Agency named it the fourth-most climate-resilient state in 2017.

MW: You're right, we are seeing climate refugees here. We also had pandemic refugees. We've seen big new pressures on our housing markets, and that's the downside. We've long had an acute housing crisis, but it's worse than it's ever been. The silver lining is that it may finally force Vermont to get serious about putting in place land use rules at the local and state level that make it possible to build more housing.

Above: Solar panels on the roof of the Laurentide Apartments. Built in 2019, the 76-unit complex includes 52 apartments designated as affordable.

Act 250, Vermont's landmark land use law, was enacted in 1970 to limit development. In 2023, the Vermont House passed the HOME bill, which would end single-family zoning and make changes to Act 250 intended to encourage the development of affordable housing.

We desperately need more housing. We've got to get better about that, and I think there will be environmental benefits if we do. To me, more people living in a green city like Burlington is a good trade-off for the environment.

AF: Are there other strategies that you have in mind for keeping or making green Burlington affordable? Burlington has a successful community land trust, you encourage accessory dwelling units, you have inclusionary zoning—what's next?

MW: We have a lot of work to do on our zoning ordinance and our statewide land use reform. Many projects in Vermont now, good projects—good green, energy-efficient projects in settled areas— have to go through both local and statewide land use permitting processes, an almost entirely redundant process that slows things down, adds a lot of costs, and creates all sorts of opportunity for obstruction. We're pursuing three major upzoning efforts right now, and there's a big conversation about Act 250 reform happening in the state as well.

AF: Finally, what advice do you have for other city leaders to take similar climate action, especially in places that aren't primed for it quite as well as Burlington is?

MW: Whenever I talk to other mayors about this, I try to make the point that this is an area where political leadership and the will of the community can have a huge impact. When I came into office, we had almost no deployed solar here in Burlington. We made it a priority. We changed some rules about permitting. We made it easier for consumers to have solar installed on their homes. The utility played a role, and over a very small number of years, we became one of the cities in the country that had the most solar per capita. We're number five in the country, the only East Coast city in the top 20 at one point, and it's not an accident. This is making a decision to lead in this area and to make change. You can have a big impact.

In the most recent Shining Cities report from Environment America, Burlington ranked seventh in the nation, with 223 watts of installed solar per capita, and was the top-ranked city in the East.

At a time when the climate emergency is an existential threat . . . mayors and cities can demonstrate on-the-ground progress. When we do that, we show everybody else what's possible.

At a time when the climate emergency is an existential threat, when the federal government is paralyzed in its ability to drive change, and when many state governments are similarly gridlocked, mayors and cities can demonstrate on-the-ground progress. When we do that, we show everybody else what's possible.

Vermont native Miro Weinberger holds a bachelor's degree in environmental studies and American studies from Yale and a master's in public policy and urban planning from Harvard's Kennedy School of Government. Early in his career, he worked on Capitol Hill; with a community development organization in Yonkers, NY; and at Habitat for Humanity in Georgia, Florida, and New York. He returned to Vermont in 2002 to cofound a development group in Burlington that created more than 200 units of affordable and market-rate housing in under a decade. He has served as mayor since 2012.

JESSE ARREGUÍN

On Center Street in downtown Berkeley, where residential development is now booming, 2016.

Jesse Arreguín was elected mayor of Berkeley, California, in 2016, becoming the first Latino to hold the office and, at 32, the second-youngest mayor in the city's history. The son and grandson of farmworkers, Arreguín grew up in San Francisco. At nine, he helped lead efforts to name a city street after activist Cesar Chavez, beginning a lifelong commitment to social justice. As mayor, he has prioritized affordable housing, infrastructure, and education in this city of 125,000. Fittingly, during this conversation focused on housing and the task of building more of it, the sounds of construction could be heard outside the fifth-floor office suite at City Hall. Arreguín announced in 2023 that he would run for state senate in 2024.

The acronym NIMBY—for "Not In My Backyard"—reflects the sentiment of residents who oppose new development in their neighborhoods, particularly multifamily or affordable housing. More recently, a prohousing YIMBY ("Yes In My Backyard") movement has emerged.

ANTHONY FLINT: It seems like Berkeley has become a national symbol of the YIMBY/NIMBY divide. What should developers be contributing to boost supply, diversify housing options, and increase density at appropriate locations?

JESSE ARREGUÍN: I think a lot needs to be done by government. We're seeing leadership demonstrated by our governor, by the state legislature, by our attorney general—who established a housing strike force to enforce state housing laws—and by regional and local government. In Berkeley, over the past several years, we've taken significant steps to pass laws to streamline production and encourage a variety of housing options in our community.

We've also made a commitment to end exclusionary zoning. One reason Berkeley is a symbol of the YIMBY/NIMBY debate is that it is the birthplace of exclusionary zoning. In 1916, Berkeley adopted its first zoning ordinance to zone the neighborhoods in the Elmwood District as single-family to prevent the construction of a dance hall. Not surprisingly, the people who would frequent that dance hall would predominantly be people of color. Sadly, single-family zoning was founded on racial exclusion.

My perspective on zoning, on housing issues, has evolved, because the crisis in Berkeley and in California has worsened significantly in the past five years. We have increasing numbers of people who are experiencing homelessness, tent encampments on our streets, working families who can't afford to live in the communities they work in, students who can't afford to live in the communities they go to school in. The status quo is not working, and we need to take bold action.

I think developers are eager to see leadership on the part of government. We need to meet them in the middle and do what we can to make it easier for them to build. At the same time, we have to make sure they're providing community benefits while we're seeing market-rate construction, particularly in communities with significant amounts of displacement and gentrification. We have historically Black neighborhoods where homes now sell for $2 million. Our Black population has declined from 20 percent in 1970 to 7 percent now. That's a direct result of decisions by the government to not build housing, and of the astronomical cost of housing in Berkeley.

Berkeley saw a 33% increase in people experiencing homelessness between 2015 and 2019, according to the nonprofit EveryOne Home. The number of unhoused residents decreased 5% from 2019 to 2022, while the countywide rate climbed by another 22%.

Below: A residential neighborhood in Berkeley, where the median sales price for a single-family home was $1.5 million in 2023.

AF: Let's talk about gentrification and real estate speculation, a problem in many cities. Los Angeles recently started a program of land banking parcels near transit stations. Is that the kind of thing that is going to be necessary when you're obviously in white-hot market conditions here?

JA: I think so, and we are prioritizing public land for affordable housing. We've converted parking lots to affordable housing projects; we have one being constructed right up the street—140 units of affordable housing and permanent supportive housing. It's the largest project we've ever built for housing the homeless. We need to prioritize public land for public good. There's no question about that.

Completed in 2022 on a former city parking lot, the affordable housing project Arreguín references consists of two buildings: Berkeley Way Apartments, which offers 89 affordable units for low- and very-low-income renters, and the adjacent Hope Center, which holds 53 permanent supportive apartments, a 32-bed homeless shelter, 12 transitional beds for veterans, and a community kitchen.

In November 2022, Berkeley voters approved Measure M, establishing an annual tax of up to $6,000 on residential properties that are left vacant for more than 182 days in a year.

I agree, we need to look at land banking. We need to provide money so nonprofit developers can buy parcels to keep them permanently affordable. We need to look at how we can support land trusts—not just buying properties, but buying buildings in order to keep them permanently affordable. That is part of Berkeley's housing strategy; it's not just new construction, but also the preservation of naturally occurring affordable housing. We need to focus on the three P's—and I say this often: *production* of new housing, *preservation* of naturally occurring affordable housing, and *protection* of existing residents from displacement.

AF: How might a vacancy tax, similar to what we see in San Francisco and Oakland, address this issue of the burgeoning value of land?

Below: A rendering of The Hub, a 26-story mixed-use student housing complex in the city's development pipeline.

JA: We recently placed on the ballot a residential vacancy tax, which is a little bit different from Oakland's; it doesn't focus on vacant parcels, but on vacant homes and residential units. There are some who have said, "Well, we have thousands of vacant units, and therefore, we don't need to build more housing." That's absurd. We need to build housing, and we also need to put housing that is off the market back on the market.

The more we can address actions by speculators and by scofflaws—I would characterize people who keep properties blighted and vacant for many years as scofflaws—the more we will address the artificial constraining of the market and put units back on the market. We spent a lot of time crafting this vacancy tax and really thought through the situations in which units could be vacant legitimately. The focus is not on small property owners but on owners of large rental properties, because part of what we are seeing is, frankly, speculation of the market.

We need to prioritize public land for public good. There's no question about that.

We hope that at some point we won't have to charge a tax, because all the housing is being rented or used. That's the goal of the vacancy tax: not to penalize, but to incentivize owners of multifamily properties to use the properties for their intended purpose. But again, this is not a panacea; this is not the solution to the housing crisis. We need to build new housing. What we have is a crisis that is decades in the making through deliberate actions on the part of government, through racial segregation or redlining, through fierce resistance to building housing, and through policies that have constrained the production of housing.

AF: As a hub of innovation, Berkeley has a thriving economy. Do you believe it's going to be possible for more workers in Berkeley to be able to live in Berkeley, or is there a built-in imbalance that you just have to manage and come to terms with?

JA: I think it's possible, but it's going to require that we build thousands and thousands of units of housing, that we prioritize building housing around our transit stations, that we look at upzoning low-density commercial neighborhoods, that we look at building multifamily housing in residential neighborhoods. Every part of our city needs to meet its responsibility to create more housing. No part of our community can be walled off to new people.

Facing page: The parking lot
at the North Berkeley BART
station, site of a proposed
development that would include
affordable housing. The plans
ignited a debate between prohousing advocates and nearby
residents concerned about
neighborhood character.

Above: The city is working to
improve pedestrian safety. Its
Vision Zero policy, adopted in
2019, aims to eliminate fatal and
severe traffic collisions by 2028,
largely through engineering and
street redesign.

That gets to the core of who we are, who we say we are as a city. Are we a city of equity and inclusivity? If we are, then we need to welcome new people into our community and create opportunities for them to live here—people who previously lived here and were displaced, people who work here but can't afford to live here. And, obviously, there's a climate benefit if people do not have to drive an hour or two hours to get to Berkeley. It reduces cars on the road, reduces greenhouse gas emissions, and helps us mitigate the impacts of climate change. Building dense, transit-oriented development is a critical part of taking bold climate action; our land use policies and our actions to encourage more dense housing are really critical climate action strategies.

AF: Could you talk about the importance of bicycle and pedestrian safety in your view of how the city functions? How is Berkeley doing in that regard?

JA: Because we have such high numbers of people who bike to work and walk and use alternative modes of transportation, we need to make it safer and easier for people to get around town. Sadly, we've seen an increasing number of collisions between cars and bicyclists, and pedestrians. Like many communities, we've adopted a Vision Zero policy that's focused on reducing traffic injuries and fatalities. We are looking at how we can redesign and reconstruct our streets to make them safer for people who walk and bike. Obviously, being the home of the University of California, we have a lot of young people, and we need to make it safer for students and for our residents to get out of their cars and to choose noncarbon-intensive modes of mobility.

AF: On climate, what else can Berkeley do? How is this region addressing the climate crisis?

JA: The best way for Berkeley to address the climate crisis is through recognizing, one, that it's not a crisis, it's an *emergency*— and we see the real material effects of it here in California. We've had some of the most devastating wildfires in California history over the last five years, and Berkeley is not immune to the threat of wildfire. That's a pretty telltale sign that the climate emergency is here, that it's not going away, and that we need to take bold action.

I'm proud that Berkeley has been a leader in combating climate change. We were one of the first cities to adopt a climate action plan. Obviously, building dense infill housing is a critical part of that. We do need to promote more electric mobility, whether it's through micromobility or through converting heavy-duty and light-duty vehicles to electric, and California's been a leader at that. While there are very ambitious targets that the state has set to transition our vehicle fleet to electric, we don't have the infrastructure to support that yet. We hope that with the new federal bipartisan infrastructure law and the climate law there will be significantly more resources available that we can leverage to expand that infrastructure in California.

Electrifying our buildings is important too, and Berkeley was the first city in California to adopt the ban on natural gas and require that newly constructed buildings be all electric. We're also looking at how we can get existing buildings to be electric, which is much tougher.

We have to adapt to climate change, whether it's through how we address wildfire risk or sea-level rise. Berkeley's along the San Francisco Bay, and we know that parts of our city, unless we do something, are going to see significant flooding and inundation. That's where the regional approach comes in. These issues can't be solved by one city. A lot of work's been done at the Metropolitan Transportation Commission and the Association of Bay Area Governments—our regional planning agency and council of governments—to bring government agencies together to explore strategies. I think that's an area where regionalism is going to make a difference.

Facing page: Wildfire smoke on the University of California, Berkeley campus in 2020 (top). The increasing threats posed by the climate crisis have led governments across the Bay Area (bottom) to collaborate.

In April 2023, a federal court ruled in favor of the California Restaurant Association, striking down Berkeley's 2019 ban on installing natural gas lines in most new construction. The court ruled that the ban was preempted by federal law. The city vowed to push back.

Jesse Arreguín was born in Fresno and grew up in San Francisco, where he and his family experienced evictions and housing insecurity that shaped his views on affordable housing. After graduating from the University of California, Berkeley—the first in his family to graduate from college—Arreguín stayed in the city, serving on boards including the Housing Advisory Commission, Rent Stabilization Board, Zoning Adjustments Board, and Planning Commission. He was elected to the city council in 2008, elected mayor in 2016, and reelected in 2020.

FREETOWN

YVONNE AK

Launching an initiative to protect vendors from extreme heat
by installing market shades, Calaba Town Market, 2022.

I-SAWYERR

FREETOWN

Yvonne Aki-Sawyerr was working in London's financial industry when a global health crisis called her home. In 2014, she returned to Freetown after two decades away to help plan and implement Sierra Leone's response to the Ebola crisis, then served as a team lead for second-phase recovery efforts. Aki-Sawyerr took office as mayor and head of the Freetown City Council in May 2018. A leader in the C40 global climate network, she launched the Transform Freetown planning initiative and appointed Africa's first chief heat officer to confront the impacts of climate change. Aki-Sawyerr was elected to a second term in June 2023.

Below: Founded in the late 1700s as a community for people who had been enslaved in England and America, Freetown is home to more than one million people.

ANTHONY FLINT: Could you talk about the Transform Freetown initiative as a planning and action framework, and your assessment of its progress?

YVONNE AKI-SAWYERR: I ran for office in 2018, motivated by concerns around the environment and sanitation. My campaign message, "for community, for progress, for Freetown," translated into Transform Freetown. It focuses on four priorities: Resilience, Human Development, Healthy Cities, and Urban Mobility.

Resilience includes environmental management; it also includes urban planning, because you cannot separate the two, and revenue organization, because sustainability will only come from the city's ability to sustain and generate revenue itself. The Healthy Cities cluster includes sanitation, which goes very closely with environmental management for Freetown and many African cities. If you think about climate change, our teeny-weeny contribution to climate change, a lot of it comes from methane from open dumping, which also has huge health implications. So the Healthy Cities category addresses sanitation, health, and water.

Having come into office with those high-level areas of concern, we created 322 focus groups with about 15,000 residents to get their views on affordability, accessibility, and availability of services across those sectors. We invited the public sector, private sector, and the international community via development partners and NGOs to participate in roundtable discussions.

Out of that process came 19 specific, measurable targets that we're working toward under Transform Freetown. We report against them every year back to the city, back to our residents. It really has been a way of introducing greater accountability, of holding our own feet to the fire, and it's very much community owned and community driven.

AF: Given all the climate threats the city faces, you appointed a chief heat officer. Why was a chief heat officer necessary, and what have been the results thus far?

YA: I'm asked often, how do you get ordinary people interested in climate change? In our case it's not hard, because the consequences of climate change are intensely felt in our parts of

In 2021, the Arsht-Rockefeller Foundation Resilience Center and the Extreme Heat Resilience Alliance established a pilot program of chief heat officers—including Eugenia Kargbo in Freetown—who are charged with leading heat risk-reduction responses in cities around the world. The other participating cities include Athens, Dhaka, Melbourne, Miami, and Santiago.

Right: Freetown installed shades over three open-air markets, providing relief for an estimated 2,300 female vendors. The low-maintenance structures, built of waterproof, semi-translucent polycarbonate and framed with steel pipes, include solar panels that power lighting, allowing vendors to extend their operations into the cooler evening hours. The city is currently raising funds for maintenance, repairs, and expansion of the project to new markets.

the world. We suffer greatly from flooding and landslides, hence my concern with the environment and being able to mitigate those impacts.

The Arsht-Rockefeller Foundation Resilience Center really got us thinking about the fact that there are more deaths from extreme heat than there are from the more visible and tangible disasters like floods and landslides. Extreme heat, particularly where water is in short supply, is a major impact of the warming climate.

In our case, the vulnerable are mainly those living in informal settlements—that's 35 percent of our city's population. In those informal settlements, the housing structures are typically made from corrugated iron, so with increased temperatures, you're effectively living in an oven. The other aspect is that we have an informal economy. Around 60 percent of women in our city are involved in trading. Most of our markets are outdoors, so they're sitting in the sun all day long. Doing that work under the intense heat means that other negative health consequences are exacerbated.

The chief heat officer has worked with market women and has gotten funding from Arsht-Rockefeller and Atlantic Council to install shade covers in three of our open-air markets. It's great to see the enthusiasm of the women; they're saying, "Are we going to get this all the way along the market? We can see where it's starting, where it stops, but we need it too."

With the chief heat officer, we'll be able to embark on some research, collecting data to identify heat islands. Anecdotally, we have a sense of where they are—mainly in the informal settlements, but potentially also in the middle of the city. We need to be able to make arguments to challenge what's going on with the lack of building permits, and land use planning being devolved to the city, and the massive deforestation that continues unabated.

I'm asked often, how do you get ordinary people interested in climate change? In our case it's not hard, because the consequences of climate change are intensely felt in our parts of the world.

AF: What are your hopes for other climate mitigation projects, including the initiative to plant a million trees? How did that come about, and how is it going?

YA: It came about because there's an appreciation that we were losing our vegetation and that worsens the effect of extreme weather events, as when heavy rains led to massive mudslides in 2017. The lack of forestation is a major part of that. Our goal is to increase vegetation cover by 50 percent.

Planting the million trees is the long-term plan, but in the meantime, you still have the runoff from the mountains filling the drains with silt. Our annual flood mitigation work identifies the worst of these areas and clears the silt so that when the rains come, the water can still flow. On a smaller scale, we've also been able to build roughly 2,000 meters (6,500 feet) of drainage in smaller communities. Beyond that, we've invested heavily in disaster management training and capacity building.

Climate change impacts are really pervasive. If people are experiencing crop failure in the rural areas outside Freetown, it will eventually drive a rural-urban migration. If people are unable to sustain their livelihoods, they're going to come to the city looking for some means of making a living.

Below: Freetown enlisted roughly 1,000 residents to plant, geotag, and track trees in their communities. Newly planted trees are verified and checked for health each quarter. The city has reached its goal of planting 1 million trees, and is now targeting 5 million by 2030 and 20 million by 2050.

Freetown's population more than doubled between 1985 and 2015, and is expected to double again in the next 20 years.

Previous page: A devastating mudslide killed hundreds of residents in 2017. City leaders are working to change land policies to help avoid such catastrophes in the future.

That pressure of population growth in the city is something else that we have to deal with—whether it's introducing the cable car to improve transportation and reduce greenhouse gas emissions, or encouraging the government to devolve land use planning and building permit functions so that we can actually introduce land management actions. These actions save lives and save property but also protect the environment and prevent people from building properties in waterways and streams and canals, which currently happens. All of this is made worse by not using legislation and urban management tools such as land use planning and building permitting in a constructive manner.

AF: Could you describe Freetown's property tax reform efforts, and the outcomes you've seen, in the overall context of municipal fiscal health?

YA: We worked on this property tax reform starting with only 37,000 properties in the database of a city—a capital city—with at least 1.2 million to 1.5 million people. When I came in, it was clear that 37,000 properties was not reflective of reality. But also, the manual system that they operated, literally with a ledger book, was not fit for purpose in the 21st century.

One of our 19 targets is to increase property tax income fivefold by 2022. To go about doing that, we secured funding and part-nerships to digitize. We changed from an area-based system to a point-based system. We worked on that by taking a satellite image of the entire city and building an algorithm to give weightings to features like roofs, windows, and location, then comparing that against a database of 3,000 properties whose values were deter-mined by real charter surveyors. We got the old-type assessment done. We were able to identify outliers and refine the model and eventually build a model which we now use as our property base.

Through that process, we moved from 37,000 properties to over 120,000 properties. That meant we were able to meet our target of increasing our property tax revenue, going from $425,000 to over $2 million. That in itself is the pathway to sustainability and being able to invest.

A big part of fiscal health is that sustainability, but unfortunately, the Ministry of Local Government halted collections while devel-oping national tax reform guidelines, so we were without revenue

for about a year. We have started collecting again, but as you can imagine, compliance levels will take a long time to recover.

Above: The northern section of downtown Freetown, with the waters of the North Atlantic visible in the distance.

AF: Where do you find inspiration in the face of so many challenges?

YA: From the fact that we have been able to make a difference in the lives of Freetonians. We've been able to test and to see how much can be achieved if one is given the space to do so. We know that so much is possible, and so we keep going.

Prior to her election as mayor of Freetown in 2018, Yvonne Aki-Sawyerr was a finance professional with over 25 years of experience in the public and private sectors; she had been involved with the campaign against blood diamonds and was instrumental in the response to the Ebola crisis in 2014. Aki-Sawyerr has delivered two TED talks, about turning dissatisfaction into action and the capital city's initiative to plant a million trees, and was named to *Time* magazine's TIME100 Next list of emerging leaders in 2021, and to the BBC's 100 Women list in 2020. She holds degrees from the London School of Economics and Freetown's Fourah Bay College.

SEOUL

OH
SE-HOON

Greeting a supporter two weeks before election day, 2021.

A lawyer by profession, Oh Se-hoon was elected in April 2021 to serve as the 38th mayor of Seoul, South Korea's capital city of almost 10 million. Oh had previously served as Seoul's mayor from 2006 to 2011; initiatives he introduced then around housing and governance—including the Seoul Hope Plus Savings Account, an asset-building program for the working poor—earned public service awards from the UN. Oh's election victory in 2021 was attributed in part to the public's dissatisfaction over housing costs, which he promised to address. In late 2022, a stampede in Seoul's Itaewon district killed 159 people and attracted global media attention; the mayor offered a tearful public apology, pledging to improve public safety.

As part of a Garden City Seoul program that aims to increase and improve accessibility to green space, the city is investing more than $140 million over five years to build 250 miles of green paths that will connect an existing 1,000 miles of urban parks and forests, creating a 1,250-mile Seoul Green Path.

ANTHONY FLINT: What is your vision for the redevelopment of the city and the creation of more meaningful public space and parks, including plans for the transformation of the former US military base at Yongsan?

OH SE-HOON: Seoul has emerged as a globally competitive metropolis thanks to urban development. In the decade leading up to 2021, the city prioritized conservation, rather than creating convenient and comfortable public spaces. Moving forward, Seoul will pursue a recreation strategy and take steps to break down barriers between conservation and development. We're redefining our approach to urban planning; our vision is to transform Seoul into an attractive, economically active city with expanded green space downtown, including the Han River, and to develop a wide range of recreational and cultural facilities. The objective is to create an "emotional city" where culture and art are integrated into people's daily lives, and nature serves as a backdrop for reflection.

Yongsan is the last piece of land in Seoul that's available for future development. It will become the political, economic, and ecological epicenter of the future Seoul, and all of Korea. After the presidential office was relocated to this area in 2022, it became the focal point of Korean politics. The former train depot site will be transformed into an international business district. The relocation of the US military base is 31 percent complete. It's difficult to pinpoint the exact date when the transfer will be completed, but the area will be transformed into hundreds of acres of green space—a place of rest and tranquility for citizens.

In April 2022, Seoul announced the Green Urban Space Recreation Strategy. The idea is to decrease the building-to-land ratio and raise the floor area ratio, easing building restrictions in the urban core. This is expected to quadruple the current ratio of urban green space from 3.7 percent to over 15 percent. We're prioritizing the revitalization of the outdated Jongmyo and Toegye-ro area (the Sewoon Shopping Center district).

Above: Visitors explore Seoul's Cheonggyecheon Stream, which was restored and transformed into a public space with the removal of a highway overpass in 2005.

A walled-off compound of roughly 500 acres in the center of Seoul, Yongsan Garrison has been off-limits to city residents for over 100 years. Established in 1904 by Japan during its occupation of Korea, the site became the headquarters for US and UN forces during the Korean War, then served as a US military base. The US Army is moving its base to a new location south of Seoul.

The objective is to create an "emotional city" where culture and art are integrated into people's daily lives, and nature serves as a backdrop for reflection.

Above: The new design for Yongsan Park will transform the former military base into an iconic public space centered on the theme of healing.

In August 2022, we unveiled the Great Sunset Han River Project, which will usher in an era of 30 million international visitors. The project aims to make the Han River a popular urban space by enhancing its allure and convenience. The plans include a mega Ferris wheel, Nodeul Art Island, and a floating performance stage. And last February, Seoul announced the Urban and Architectural Design Innovation initiative, which aims to increase the city's competitiveness through innovatively designed buildings. Business plans will prioritize design elements to encourage creative public building design.

AF: You have said there needs to be a better range of housing options, particularly for young individual renters. How are you addressing the problem of housing affordability?

OS: Housing problems prevent individuals from climbing the social ladder. Housing is the most expensive component of essentials such as food, clothing, and shelter, and is becoming a source of pain and anxiety for citizens, particularly young people. According to a Seoul Metropolitan Government survey, jeonse loans for young people have increased sixfold in the last four years, and 59.4 percent of young single-person households live in rental housing.

Seoul is pursuing various housing and housing support policies to help young people participate in social and economic activities without having to worry about finding a place to live, including providing public housing, improving the quality of rental housing, and providing private youth housing at below-market rates to help young people accumulate assets and start their own families.

Generation-integrated housing, which can house parents, children, and grandchildren, can help address daily challenges and social issues such as rapid aging and childcare. We also intend to provide senior-friendly public housing with residential, medical, and convenience amenities. The government's ultimate objective is to stabilize home prices.

AF: What are the key elements of Seoul's current climate action plan, and how do you envision that being a model for other cities?

OS: In response to the climate crisis, the Seoul Metropolitan Government established the 2050 Seoul climate action plan to achieve carbon neutrality by 2050. It was submitted to C40 and received final approval in June 2021. The plan aims to create a sustainable city where people, nature, and the future coexist. It outlines policy goals in five major areas: build and retrofit one million low-carbon buildings by 2026; expand electric vehicle supply to 400,000 units and install EV chargers by 2026; provide various renewable energy sources (such as fuel cells, geothermal, hydrothermal, and solar); reduce waste, promote recycling, and prohibit direct landfilling; and expand urban parks and forests to mitigate greenhouse gas emissions and enhance urban resiliency.

South Korea's jeonse system, widely used since the 1960s, is a unique housing lease arrangement in which tenants pay a large deposit upfront—as much as 70% of a home's value, thus making a loan necessary in most cases. The money is returned to them in full at the end of the lease term, which is typically two years.

The Seoul Metropolitan Government is investing almost $8 billion over five years to advance its climate action plan, which is expected to create 70,000 jobs by 2026 and reduce carbon emissions by 35 million tons annually by 2050.

Aiming to get 400,000 gas-powered vehicles off the streets by 2026, the city will require all new vehicles to be electric beginning in 2025.

Our goal is to reduce greenhouse gas emissions by 30 percent, compared to 2005 levels, by 2026. It will take a concerted effort on a global scale to solve the climate crisis. Seoul will share its best practices with mayors of cities worldwide and engage in dialogue with them to combat the climate crisis.

AF: Tell us about how Seoul has become a smart city, including through the use of robotics and apps, and your exploration into virtual reality.

OS: Seoul is a global smart city that has been an outstanding leader in fields like e-government—we've been named the best e-government seven times. Currently, 16 self-driving vehicles are on the road at all times in four areas: Sangam, Gangnam, Cheonggyecheon, and the Blue House (Gyeongbokgung Palace, the former presidential residence). We plan to offer autonomous vehicle service across the city by 2026 and to become a global standard model city for autonomous driving.

The city added a fifth autonomous vehicle zone, Yeuido, in the summer of 2023.

Above: Oh introduces Metaverse Seoul.

We're also using robots and AI technologies across our public administration. Robo Manager, our robotic public servant, handles simple administrative tasks, such as the delivery of documents. Assistant Manager Seouri, a virtual public official and internal chatbot, helps employees with complex business procedures. Metaverse Seoul was named one of the best inventions of 2022 by *Time* magazine; it was the only public sector invention on the list. Metaverse Seoul is a virtual place where anyone can enjoy Seoul, since it's not limited in time or space and does not discriminate in terms of gender, disability, or occupation. Seoul intends to implement the metaverse ecosystem across all of its administrative services, including the economy, culture, tourism, and citizen complaints.

In collaboration with the World Smart Cities Organization, Seoul recently established the Seoul Smart City Prize. The prize is intended to promote Seoul's core values as well as to discover inclusive and innovative projects to share with the world.

AF: You have traveled to South America and Africa to talk about city administration. What did you tell them about managing the modern city?

OS: I traveled to Lima, Peru, and Kigali, Rwanda, several years ago as part of a Korea International Cooperation Agency advisory group. Lima was highly interested in Seoul. I discussed my experiences with the Han River Renaissance Project and housing. I also discussed the Women Friendly City Project, which aimed to implement women-friendly facilities, including pedestrian roads, parks, restrooms, housing, and public transportation. I went to the sites where Lima's major projects, such as the Rimac River Project and the Costa Verde Project, were being carried out. And I organized a seminar to examine housing policies including site development and rental policy.

At the time of my visit to Kigali, in 2014, the city was still working hard to heal the wounds left by the atrocious genocide that had killed one million people 20 years prior. I was impressed by how they were overcoming the tragic history, declaring *Kwibuka*, "let us remember," rather than seeking vengeance. I admired how they

Above: Oh introduced the Han River Renaissance Project, a precursor to today's Great Sunset Han River initiative, during his first term as mayor in 2007. The city built bike paths, pedestrian bridges, floating islands, and other infrastructure with the goal of increasing open space and recreational opportunities.

transformed their hatred into reconciliation. Urban reconstruction is a major concern in Rwanda, so I passed on my experience in urban planning, housing, and tourism—especially the importance and growth potential of tourism. From Peru to Rwanda, during overseas advisory activities and volunteering, I learned first-hand how you learn as you teach, and you receive as you give. It reminded me of how important it is for a leader to be inclusive and reconciliatory.

AF: What is your view of land value capture in private real estate development, and how it can be used to finance infrastructure, housing, and other needs?

OS: In exchange for infrastructure during private real estate development, the Seoul Metropolitan Government provides floor-area-ratio incentives. Through this exchange, the government may acquire infrastructure such as roads and parks and essential community amenities such as libraries, childcare facilities, cultural facilities, and youth facilities, as well as public rental housing and public rental industrial facilities. Between August 2015 and January 2023, these policy incentives yielded 357 public contribution facilities equivalent to approximately $5 billion. Furthermore, the revised National Land Planning Act, which went into effect in July 2021, allows for both in-kind items such as facilities and cash payments that can be used throughout Seoul. The Seoul Metropolitan Government will use these funds to cover operating

expenses for essential facilities, the expansion of roads and rail-ways, and new transportation projects.

The current zoning system will be revamped to maximize land efficiency in underutilized spaces. It will pursue two pillars of urban competitiveness: integrating residential and commercial uses and expanding urban green space. We're abolishing the rigid 35-floor regulation on residential buildings that acted as a headwind against change, easing building regulations such as height and floor area ratio that impeded urban center devel-opment. We're also expanding parks and green areas.

Seoul is reinventing itself in ways other than just modifying its urban planning practices. With the city's attractiveness in mind, the Seoul Metropolitan Government comprehensively considers factors that significantly impact a person's happiness, such as leisure, health, safety, and environment, as it builds the city.

Born in Seoul, Oh Se-hoon studied at Korea University, graduated from Korea Univer-sity's School of Law, and was a fellow at the Graduate School of Social Science and Public Policy at King's College London, where he focused on job creation and economic growth in major cities around the world. Oh cofounded the Korea Federation for Environ-mental Movements and, as a lawyer, helped to establish the right to sunlight for the first time in South Korea's history. He was a member of the National Assembly of South Korea from 2000 to 2004, served as mayor of Seoul from 2006 to 2011, and returned to office in 2021.

AFTAB PUR

EVAL

With chief of staff Keizayla Fambro after narrating a tour of East Westwood, part of a video series visiting each of Cincinnati's 52 neighborhoods, 2022.

The population of Cincinnati is growing after years of decline, and companies are increasingly interested in putting down roots in the city thanks to its strategic location. There's even talk of southwestern Ohio becoming a climate haven. But any resurgence in a postindustrial legacy city comes with down-sides, including the potential displacement of established residents and the rapid disappearance of affordable housing options. Mayor Aftab Pureval, elected in 2021, says affordability and displacement are his biggest concerns as Cincinnati—currently home to nearly 310,000 people—gets a fresh look as a desirable location. Pureval has made stimulating equitable economic growth in the city's 52 neighborhoods a top priority, along with reforming and improving public safety and zoning, increasing affordable housing, and building climate resilience.

ANTHONY FLINT: You've attracted a lot of attention for what some have called a "heroic undertaking" to preserve the city's single-family housing stock and keep it out of the hands of out-side investors. Briefly, walk us through what was accomplished in coordination with the Port of Cincinnati.

AFTAB PUREVAL: To provide a little more context, Cincinnati is a legacy city. We have a proud, long tradition of being the final destination of the Underground Railroad. We were the doorstep to freedom for so many slaves who were escaping that horrific experience. We have a lot of historic neighborhoods, a lot of historic buildings, and a lot of aging infrastructure and aging single-family homes, which—paired with the fact that we are an incredibly affordable city in the national context—makes us a prime target for institutional investors.

In July 2022, national brokerage Redfin reported that median rents in Cincinnati had soared 39% year-over-year, more than in any other US city. In May 2023, in a market of slower overall growth, the increase in Cincinnati's median observed rent still led the nation at 7.9% year-over-year, according to Zillow.

Unfortunately, Cincinnati is on national list after national list about the rate of increase of our rents. It's primarily being driven by these out-of-town investors—who have no interest, frankly, in the well-being of Cincinnati or their tenants—buying up cheap single-family homes, not doing anything to improve them, but

overnight doubling or tripling the rents, which is pricing out a lot of our communities, particularly our vulnerable, impoverished communities.

The city is doing a lot through litigation, through code enforcement. In fact, we sued two of our largest institutional investors to let them know that we're not playing around. If you're going to exercise predatory behavior in our community, we're not going to stand for it.

We've also done things on the front end to prevent this from happening by partnering with the Port. When several properties went up for sale because an institutional investor put them on the selling block, the Port spent $14.5 million to buy 194 single-family homes, outbidding 13 other institutional investors.

Over the past year, the Port has been working to bring those properties into compliance, dealing with the various code violations that the investor left behind, and, once they're fixed up, pairing these homes with qualified buyers. Oftentimes these are folks living in poverty, or lower-middle-class families who've never owned a home before.

Above: A Main Street mural of singer James Brown, whose recording career began in Cincinnati in the 1950s, offers a glimpse of the city's vibrant history.

The Port of Greater Cincinnati Development Authority is a public agency that partners with municipalities, foundations, economic development organizations, and others on residential and industrial development and revitalization projects.

Below: One of more than 190
homes purchased by the Port
of Cincinnati as part of an effort
to fend off speculators and
increase local homeownership.

This year we're making three of those 194 homes available for sale. It's a huge success across the board, but it's just one tool that the Port and the city are working on to increase affordability of housing in all our neighborhoods.

AF: What have you learned from this process that might be transferable to other cities? It takes a lot of capital to outbid an institutional investor.

AP: It does require a lot of funds. That's why we need more flexibility from the federal government and the state government to provide municipalities with the tools to prevent this from happening in the first place. Now, once an institutional investor gets their claws into a community, there's very little that the city can do to hold them accountable.

The American Rescue Plan Act (ARPA) provided roughly $280 million to Cincinnati. In 2022, Cincinnati allocated $2.5 million in ARPA funds to the Port to help construct or renovate approximately 40 workforce or low-income housing units by the end of 2024.

The better strategy, as we've seen this time, is to buy up properties on the front end. A lot of cities have a lot of dollars from the federal government through the American Rescue Plan Act (ARPA). We have used a lot of ARPA dollars not just to get money into the hands of people who need it most, which is

critically important, but also to partner with other private-public partnerships or the Port—to give them the resources to buy up the land and hold it.

This is a unique time where cities have more flexibility with how they use resources coming from the federal government. I would encourage any mayor, any council, to think critically about using these funds not just in the short term, but also in the long term to address some of these macroeconomic forces.

Above: Newer homes in the Mount Adams neighborhood.

AF: Cincinnati has become a more popular place to live, and the population has increased slightly after years of decline. Do you consider Cincinnati a pandemic or climate haven? What are the implications of that growth?

We're living through a paradigm shift . . . the way we live, work, and play is completely changing.

AP: What I love about my job as mayor is my focus isn't necessarily on the next two or four years, but the next 100 years. Right now, we're living through a paradigm shift because of the pandemic; the way we live, work, and play is completely changing. Remote work is altering our economic lifestyle throughout the entire country, but particularly here in the Midwest.

I am convinced that because of climate change, because of the rising cost of living on the coasts, there will be an inward migration. I don't know if it's going to be in the next 50 or 75 years, but it will happen. We're already seeing large businesses making decisions based on climate change. Just two hours north of Cincinnati, Intel is making a $200 billion investment to create the largest semiconductor plant in the country. Two of the reasons they chose to build it just north of Cincinnati are access to fresh water—the Ohio River in the south and the Great Lakes in the north—and our region's climate resiliency.

In 2023, Moody's ranked Cincinnati among the 10 US cities least at risk from heat, drought, and sea-level rise, making it an attractive refuge for climate migrants from the coasts.

We're all affected by climate change, but in Ohio we're not seeing the wildfires, the droughts, the hurricanes, the earthquakes, the coastal erosion that we're seeing in other parts of the country, which makes us a climate-change safe haven, not just for business investment but also for people. Cincinnati is partly growing because our economy is on fire, but I believe we're going to see exponential growth over the next few decades because of these massive factors pushing people into the middle of the country.

It's challenging to make investments right now to help our legacy communities and legacy residents stay in their homes, to make Cincinnati an affordable place for them, while also keeping in mind these future residents. While Cincinnati is very affordable in the national context, it's not affordable for all Cincinnati residents because our housing supply has not kept up with population growth and our incomes have not kept up with housing prices. To make sure future investments and population growth don't displace our current residents, we've got to stabilize our market now and be prepared for that growth.

What I love about my job as mayor is my focus isn't necessarily on the next two or four years, but the next 100 years.

AF: What are the land use changes and transportation improvements that you're concentrating on accordingly?

AP: The third rail of local politics is zoning. If we're going to get this right, then we have to have a comprehensive review and reform of our land use policies. For over a year now, we've been having meetings with stakeholders to explore what a modern Cincinnati looks like. I believe it looks like a dense, diverse neighborhood that's walkable, with good public transportation and investments in public art. But right now, the City of Cincinnati's zoning is not encouraging those kinds of neighborhoods. Close

to 70 percent of our city is zoned for single-family use exclusively, which is putting an artificial cap on the amount of supply we can create. This is increasing rents and property taxes, which is causing a lot of our legacy residents, even those who own their homes, to be displaced.

If we're serious about deconcentrating poverty and desegregating our city, then we've got to look at multifamily unit prohibitions. We've got to look at parking requirements for both businesses and homes; at transit-oriented development along our bus rapid transit lines; at creative opportunities to use auxiliary dwelling units. None of this is easy.

I am confident we can make some substantive changes to our zoning code to encourage more affordability, more public transportation, and just become a greener city.

On that note, we have made a commitment that we will only buy city vehicles that are electric vehicles when they become available. We have the largest city-led solar farm in the entire country.

AF: A little bit of this is back to the future, because the city once had streetcars. Do you have the sense that there's an appreciation for that, that the city actually functioned better in those times?

Above: Downtown Cincinnati in 1951.

In June 2023, the Cincinnati City Council voted 8-0 to legalize accessory dwelling units (also known as ADUs, or auxiliary dwelling units) on any single-family lot, making it the first city in Ohio to do so. The rule includes an owner-occupancy requirement.

With more than 310,000 solar panels, the 100-megawatt New Market Solar field, built on 890 acres of former farmland east of Cincinnati, produces enough electricity to power all of the city's municipal buildings and is projected to eliminate 158,000 tons of carbon emissions annually.

AP: The city used to be dense; it used to have incredible street-cars, public transportation, and then, unfortunately, cities—not just Cincinnati but across the country—saw a steady decline of population as they started losing folks to the suburbs. Now people want to come back into the city, but we have the hard work of undoing what many cities tried to do, which was to create suburban neighborhoods within the city to attract those suburban people back. It's a little bit like undoing the past while also focusing on what used to exist.

AF: What worries you most about this kind of transition, and what do you identify as the major issues facing lower-income and communities of color in Cincinnati?

AP: Displacement. If we cannot be a city that our current residents can afford, they will leave, which hurts everything. If the city is not growing, then a city our size, located where we are in the country, we are dying, and we are dying quickly.

Cities our size have to grow, and in order to grow, not only do we need to recruit talent, but we need to preserve the families and the legacy communities that were here in the first place.

No city in the country has figured out a way to grow without displacement. The market factors, the economic factors, are so profound and so hard to influence, and the city's resources are so limited, it's really difficult. Oftentimes, I get frustrated that I don't have enough resources, enough authority, to make a meaningful impact on the macroeconomic forces that are coming into the city. Because if we get our dream, which is more investment, more growth, that comes with negative consequences, and it's difficult to manage both.

AF: Finally, back to climate change, the mayor's website says Cincinnati is well positioned to be a leader in climate change at home and abroad. What do you think the city has to offer that's distinctive in terms of climate action?

AP: All of our policy initiatives are looked at through two lenses. The first is racial equity and the second is climate. Everything that we do, whether it's our urban forestry assessment, looking at a heat map of our city and investing in trees to not just clean the

air but also cool our neighborhoods, or our investments in biochar. We are one of only seven cities in the entire world that received a huge grant from the Bloomberg Philanthropies to continue to innovate in the world of biochar, which is a byproduct of burning wood and is an incredible carbon magnet.

Ultimately, businesses and people who are looking to the future consider climate change in that future. If you're looking for a city that is climate resilient but also making massive investments in climate technology, then Cincinnati is that destination.

Biochar, a charcoal-like byproduct of burning plant waste in a low-oxygen environment, has long been used as a soil amendment to help retain nutrients—but it's gained attention for its carbon-sequestering abilities. Cincinnati's Parks Department is investing $1.1 million, including $400,000 in grant funding, to develop its own biochar facility.

Above: Distribution of free trees through the Fall ReLeaf program in 2020. The annual program prioritizes neighborhoods where tree canopy coverage is 40% or less.

Aftab Pureval was raised in southwest Ohio, the son of first-generation Americans. After graduating from the Ohio State University and the University of Cincinnati Law School, Pureval held several legal positions, including as counsel at Procter & Gamble, before entering public service. He served as Hamilton County Clerk of Courts from 2017 to 2021—the first Democrat to hold that office in over 100 years—and made a bid for Congress in 2018. Pureval was elected to his current office in 2021, making history as the first Asian American mayor of Cincinnati and of any major city in the Midwest.

SHELLY OBEROI

Speaking to an audience as part of a mega parent-teacher
meeting held at 2,500 Delhi schools, West Patel Nagar, 2023.

With a population of nearly 33 million residents and growing, Delhi is the second-largest metropolitan area in the world after Tokyo, and on track to become number one. Shelly Oberoi, 39, is mayor of the Municipal Corporation of Delhi (MCD), a governing body representing some 20 million of those people. Born in the capital city, Oberoi was named a vice president of the women's wing of the anti-corruption Aam Aadmi Party before becoming a ward city councilor in 2022. Oberoi, who had to run for the mayoral post several times due to parliamentary voting challenges, promised that "Delhi will be cleaned and transformed" in her tenure. The city has been confronting worsening pollution, tensions between Hindus and Muslims, and—like so many cities in India—strains on its natural and economic resources due to rapid population growth.

The Municipal Corporation of Delhi (MCD) was established in 1958. In 2012, it was divided into three separate municipal corporations—North Delhi MC, South Delhi MC, and East Delhi MC—in an effort to better serve the fast-growing population and vast geographical spread. To improve transparency, governance, and efficiency, MCD was unified in 2022.

ANTHONY FLINT: You're the first mayor in a decade to oversee all of central city Delhi, after reunification of the Municipal Corporation there. What kind of governing challenges and opportunities come along with that?

SHELLY OBEROI: Governing the Municipal Corporation after its unification has come along with a fair share of challenges and opportunities. On one hand, centralization of powers allows for streamlined decision-making, enhanced accountability, and improved collaboration across departments. While centralization allows for more efficiency, it also requires careful planning to ensure equitable distribution of resources to address the diverse needs of different areas within Delhi. Balancing these needs and optimizing resource allocation is a significant challenge that we are addressing at the moment. On the other hand, unification has also offered us an opportunity for policy alignment. With a unified municipal corporation, we can now align policies and regulations across all areas of Delhi. Policy alignment allows us to address issues such as education, property tax, and new initiatives in a coordinated manner, leading to more effective civil planning

Above: Paharganj Market, New Delhi.

and development across the city. This enables consistent implementation of rules and regulations, creating a level playing field and ensuring fairness and transparency in governance.

AF: You said upon being elected that you would work "to make Delhi the city that it should have been"—what does that vision look like, and what are the biggest obstacles to achieving it?

SO: My vision for Delhi is based upon the Aam Aadmi Party's 10 guarantees, as announced by our National Convenor and Chief Minister Arvind Kejriwal. These guarantees reflect the aspirations of the people and prioritize the overall well-being of the city. We have envisioned a clean and beautiful Delhi, free from the blight of landfills, where waste management systems are streamlined and cleanliness is promoted throughout the city. We are establishing a culture of transparency and accountability, ensuring a corruption-free Municipal Corporation of Delhi. Our vision also includes providing a permanent solution to the problem of parking through efficient management systems and addressing the issue of stray animals with compassionate and sustainable measures. Moreover, we aim to have well-maintained roads that prioritize safety and smooth traffic flow, improving the overall commuting experience for residents.

Each of India's 28 states and three of its eight union territories has a governor who is appointed by the president of India and chief ministers who are elected by the people and have executive authority. Arvind Kejriwal has served as chief minister of Delhi since 2015.

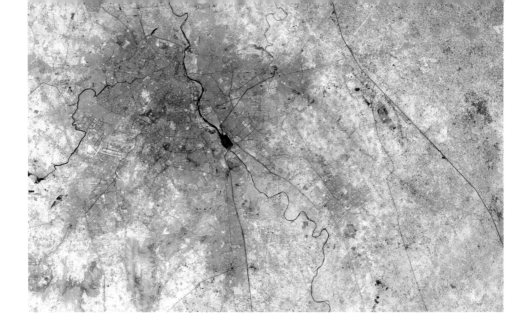

Above: Delhi more than doubled in geographic size between 1989 and 2020. This NASA Earth Observatory map shows the city's reach in 2018.

Previous page: Leaders are working to address Delhi's notorious air pollution, especially the fine particulates known as PM2.5, which cause an array of health problems.

Oberoi's party intends to make Delhi "the city of parks," adding thousands of new green spaces. In 2022, Delhi planted more than four million trees and shrubs, with plans to plant another five million saplings in 2023. Meanwhile, the public works department has transplanted thousands of mature trees to help cool the city streets.

The work of the Aam Aadmi Party's state government in Delhi is already talked about globally, particularly in the fields of education and healthcare. Chief Minister Kejriwal has administered revolutions in the landscape of India's public education and public health sectors. People have started believing that government facilities can be trusted, that they can offer them the equal standard of services for free that private facilities do at exorbitant prices.

Building on this momentum, we are working with a special focus on transforming schools and hospitals into centers of excellence. We are also enhancing parks across the city, creating green spaces for citizens to enjoy. In a welcome change, we are ensuring regular salaries for workers and offering them a better environment within the MCD to promote job security and build a motivated workforce. Simplifying the process of obtaining licenses for traders, creating a welcoming business environment, and establishing designated vending zones for street vendors are also part of our vision.

However, we acknowledge the challenges posed by rapid urbanization, budgetary constraints, stakeholder engagement, and coordination among different agencies. By recognizing these challenges and proactively addressing them, we can work toward making Delhi the city it should have always been—a thriving, inclusive, and sustainable metropolis that residents can be proud to call home and, above all, the number-one capital of the world.

AF: Regarding air quality—brought to international attention by such documentary films as *All that Breathes*—what are some short-term solutions? Please also comment on your approach regarding garbage and landfills. The two issues are related, in that the new waste-to-energy plant will seemingly help solve one problem while further contributing to air pollution.

SO: Air quality is indeed a pressing concern for Delhi, and addressing it requires a multifaceted approach that incorporates both short- and long-term solutions. However, air doesn't belong to any one geographical boundary; a lot of factors that arise in our neighboring states adversely impact Delhi. Thus, the challenge needs a concerted and coordinated approach from all stakeholders, including the central government and neighboring state governments.

Delhi has four waste-to-energy plants that convert municipal solid waste into electricity.

Air doesn't belong to any one geographical boundary; a lot of factors that arise in our neighboring states adversely impact Delhi.

The Delhi government is leading an extensive effort to reduce air pollution through its Summer and Winter Action Plans. The government accordingly decides upon short- and long-term solutions as part of these action plans, be it stopping dust pollution and industrial pollution, improving on solid waste management, or conducting real-time source apportionment studies. Under these action plans, the MCD has been delegated the responsibility of keeping a check on the factors under its domain and maintaining vigils on smaller roads under its domain. The state government regularly convenes review meetings and the MCD has extended its unconditional support to help with these efforts. Due to these efforts, the air pollution levels in Delhi have already seen a welcome change.

Under the Summer and Winter Action Plans, Delhi has taken steps including banning open burning and firecrackers, deploying hundreds of street sprinklers to combat dust pollution, and sending compliance teams to patrol the city. It has also launched two apps, Green Delhi and MCD 311, to make it easier for citizens to report concerns about pollution, illegal dumping, and other issues.

The government is aiming to clear Delhi's three major landfills—Okhla, Bhalswa, and Ghazipur—by the end of 2024, through a process of recycling, reuse, and waste-to-heat combustion. The landfills, which cover a collective 200 acres and stand as high as 150 feet, contain more than 20 million tons of waste. The MCD is processing and clearing 13,000 to 16,000 tons of waste per day.

As for garbage and landfills, we are actively working upon improving the city's solid waste management system by means of promoting waste segregation, installing Fixed Compactor Transfer Stations (FCTS), and shutting down neighborhood garbage dump yards. We have also set a plan to eliminate the three garbage landfills of the city. We are on track to completely clear off the Okhla landfill by the end of this year and the Bhalswa landfill by the first half of next year. These targets have been set by the state as part of a dedicated approach to clean the city, and Chief Minister Kejriwal has been monitoring the daily progress to further strengthen MCD's resolve toward this mission.

AF: Are there any policies in the works to address the city's notorious traffic congestion? How does that fit in with your overall plan to enhance infrastructure and make the city more resilient?

SO: Traffic as a subject is mostly beyond the domain of the MCD. In Delhi, the municipal body only looks after minor roads and neighborhood lanes, whose upkeep we've been working upon with utmost commitment ever since taking over the reins. Along with the help of our councilors and local citizens, we've been identifying all such roads and lanes that need any sort of repair and ensuring that the task is dealt with. At the larger level, the Delhi Government's Public Works Department and Transport Department are doing a great job of reducing traffic congestion in the city by upgrading the existing infrastructure, building new flyovers and underpasses, and introducing electric buses.

AF: The Delhi metro area—with a population of nearly 33 million and growing by nearly 3 percent per year—seems to warrant a more centralized form of governance. Is there any chance of reform to allow mayors in India to manage their cities as leaders do in major cities in other parts of the world?

SO: In principle, I do recognize the need for reforms that empower city leaders to effectively manage their cities, similar to the governance models observed in major cities around the world. However, the current governance structure in India has limits that we respect, and we prefer to mull about within our own landscape. In theory there is always a chance for reform and exploration of alternative models. We can explore enhancing the capacity of mayors and

local authorities through training programs, knowledge sharing, and collaboration with international city management institutions that can equip them with the necessary skills and expertise to effectively lead and manage their cities. We can also promote collaborative governance models that involve active participation of citizens, civil society organizations, and other stakeholders to facilitate better decision-making and ensure that the diverse interests and concerns of the city's residents are adequately represented.

Above: Students raise awareness of tree-planting campaigns in New Delhi. With Delhi on track to become the world's largest metropolitan area, leaders are working to involve more residents in decisions that will affect their shared future.

Born in Delhi, Shelly Oberoi was elected mayor of the Municipal Corporation of Delhi (MCD) in 2023. She studied at the Indian Institute of Management and has a bachelor's in commerce from Janki Devi Memorial College, a master's in commerce from Himachal Pradesh University, and a PhD in philosophy from Indira Gandhi National Open University. Oberoi has been an assistant professor at Delhi University and Mumbai's Narsee Monjee Institute of Management Studies, and has authored research on corporate social responsibility, global finance, and other topics.

AFTERWORD

ANGELA D. BROOKS, FAICP
President, American Planning Association

AS A CHILD, I OBSESSED OVER THE DIFFERENT NEIGHBORHOODS in my hometown of Seattle. I often wondered what was located where, how people got to places, and what fun things there were to do in different areas. I also loved summers in the South with my grandparents. I picked dinner vegetables from the field, got eggs fresh from the chicken coop for breakfast, knew everyone in the town, and rode my bike down country roads without a care.

I loved thinking about what made a place someone's home—and I also wanted to understand why some people did not *have* a home. I didn't know then that I was laying the groundwork for my future career as a planner, and for my commitment to ensuring that everyone has a safe, decent, affordable place to call home.

Today I have the honor of leading the largest planning organization in the world, the American Planning Association (APA), in an era when our cities are reimagining what they will become. In the wake of the COVID pandemic and the racial awakening sparked by the murder of George Floyd, our citizens are more engaged than ever and are looking to create spaces and places they can call home. And elected leaders, planners, entrepreneurs, funders, and others are coming together to imagine a different kind of future.

The mayors profiled in this book, along with many other mayors around the world, are working to create climate-resilient, equitable cities where all citizens can grow and thrive. They are approaching intertwined issues like housing affordability, climate change, economic development, green building, public transit, revitalization, and gentrification with new energy and new ideas. As Washington, DC Mayor Muriel Bowser reminds us in these pages, "Cities are incubators for innovation, and while we don't always have the same challenges—for example, some cities have a lot of people and not enough housing, others have a lot of housing and not enough people—we are constantly learning from each other."

This book provides opportunities for us to learn from each other. It also provides concrete examples of the kinds of collaborations that are changing our cities for the better. I serve as cochair of the Housing Supply Accelerator, a partnership between APA and the National League of Cities that is working to remove barriers to increasing housing supply in the United States. Through this initiative, planners and elected officials—including Cincinnati Mayor Aftab Pureval, an inaugural member of the accelerator steering committee who shares his insights on housing and equity in this collection—are working together with builders, financial institutions, housing policy associations, and state and federal partners to come up with solutions for creating housing in our communities. Together we will develop implementable strategies to increase housing supply and address the affordability crisis. This campaign is just one example of efforts around the world that are making cities safer, healthier, more affordable places to call home.

The American architect Hugh Newell Jacobsen once said, "When you look at a city, it's like reading the hopes, aspirations, and pride of everyone who built it." I love that idea, but I would change "everyone who built it" to "everyone who *is building it.*" After all, the work of building cities is never done. Reading about the steps mayors and public agencies are taking to address some of today's most significant urban challenges and turn them into opportunities gives me hope, and I'm inspired by the endless possibilities ahead.

The future of our cities—of the places we call home—depends on planners, elected leaders, community members, government entities, and philanthropic partners working together to create an equitable future and opportunities for all to live, grow, and thrive. It is up to all of us.

Angela D. Brooks, FAICP, is the director of the Illinois office of the Corporation for Supportive Housing and president of the American Planning Association. She currently serves on the Chicago Board of Zoning Appeals and the Illinois Affordable Housing Advisory Commission, and is cochair of the national Housing Supply Accelerator. Brooks is a native of Seattle and a graduate of Jackson State University, where she received her bachelor's degree in urban studies, and the University of New Orleans, where she received a master's degree in urban and regional planning.

AUTHOR'S NOTE

IT HAS BEEN MY DISTINCT PLEASURE to interview these 20 mayors of cities from around the world. The art of the Q&A is something I've developed over many years as a journalist, and here the conversations have been a mix of the curated and the spontaneous. The questions flowed from the work of the Lincoln Institute of Land Policy, which happily overlaps with many of the top concerns of cities—land use, zoning, housing, transportation, spatial equity, sustainability, and climate change. And indeed that was the idea for this series, conceived by veteran editor Maureen Clarke as a column for *Land Lines* magazine: to understand how innovative and energetic leaders were taking on these and other extraordinary challenges, on the ground.

These chief executives never disappointed; if the mayoralty is part proving ground, it's also been gratifying to watch many of them demonstrate their talents over time. Often, I felt like I was connecting with political rising stars at a particular moment in a promising trajectory. It wouldn't surprise me at all if one or more of these individuals ended up as president or prime minister. Or prominent cabinet member, as was the case with Boston Mayor Marty Walsh, recruited to be secretary of labor by President Biden.

One additional benefit of being a witness to history was learning about the different ways the work of the Lincoln Institute, as an international think tank, was marbled into the experience of several municipalities. The mayor of Bogotá, Claudia López, spoke of being surrounded by advisors who had been trained in the organization's Latin America program; outgoing Cleveland Mayor Frank Jackson, while laying out a vision for the equitable regeneration of that legacy city, reflected on Henry George, the author and political economist whose work inspired the organization's founder, John C. Lincoln. Phoenix Mayor Kate Gallego noted the impact of Joan Lincoln (the wife of John C. Lincoln's son, David), who served as mayor of Paradise Valley, Arizona, from 1984 to 1986 and was the first woman to hold that title. "She really was an amazing pioneer and she made it possible for candidates like me to not be anything unusual," Gallego said.

Those are some of the reasons that *Mayor's Desk* truly became a labor of love. I'd like to thank all those involved in the evolution of this journalistic endeavor, especially *Land Lines* editor Katharine Wroth, who worked with me to select candidates, craft questions, choose illustrations for the magazine, and ensure pristine copy, and editor Jennifer Sigler, who initiated and led the effort to publish this collection as a book, bringing well-honed sensibilities of art and text to the project. Additional thanks go to Jon Gorey and Amy Finch for their instrumental editorial support, and to Studio Rainwater for their thoughtful and invigorating design.

I'd also like to thank Michael Bloomberg and Angela Brooks for contributing their reflections to this collection and sharing their thoughts on the broader work of shaping our cities; the press officers and communications liaisons who facilitated these conversations in time zones near and far; and of course, the mayors themselves, for being affable and generous with their time, and, as we hope readers appreciate, for honesty and insight that promises to be timeless.

Anthony Flint
Cambridge, Massachusetts

STRATEGIES AND SOLUTIONS

Cities around the world are working to increase community engagement, strengthen fiscal and public health, fund critical infrastructure, and build more resilient, sustainable futures. The following list highlights one strategy from each of the 20 cities in this book. In some cases, the cities pioneered these programs; in others, they adapted approaches that have been used elsewhere to meet their own needs. Every example includes replicable aspects that can work in other places, and a link to learn more.

INCUBATING INNOVATION
Syracuse

In this postindustrial city, the unmanned aerial industry is taking off. As the former headquarters of General Electric and other high-tech heavyweights, Syracuse has created a welcoming environment for 21st-century companies pursuing drone research and development, through steps such as converting a vacant downtown parking garage into a tech incubator space and working with partners in the region to secure millions in state funding to develop an industry hub.

www.thetechgarden.com

BUILDING SMALLER
San Isidro, Lima

In an ongoing bid to attract younger residents, San Isidro changed its minimum required size for apartments from 200 square meters—roughly 2,100 square feet—to 45 square meters. Known for being the most expensive area of the city, the district also stopped requiring multibedroom apartments and started encouraging more residential development downtown.

msi.gob.pe/portal/plan-urbano-distrital/ plan-urbano-distrital-de-san-isidro-2023-2033

MAKING TRANSIT ACCESSIBLE
Warsaw

With one of the highest public transit ridership rates in Europe, Warsaw has taken steps to cultivate loyalty among its passengers, offering free and reduced fares for children, students of all ages, retirees, veterans, people with disabilities, and even those who can document that they were anti-Communist activists or politically oppressed. The city is introducing a clean transport zone in 2024 and phasing out diesel cars.

www.wtp.waw.pl/en/discount-entitlements

NURTURING NEIGHBORHOOD WELLNESS
Santa Monica

Big change starts with small steps—that's why the City of Santa Monica launched a microgrant program to support residents working to improve the community. Each year, the program—now managed by the Civic Wellbeing Partners coalition and funded by supporters including Cedars-Sinai and United Way—provides grants of up to $500 for projects ranging from exercise classes to financial literacy workshops.

www.wellbeingmicrogrants.org

RESEARCHING RESILIENCE
Bristol

The first city in the UK to declare a climate emergency, Bristol has undertaken a comprehensive, citizen-powered climate planning effort. The city also hosts a program called REPLICATE, in partnership with local universities and with support from the European Commission, to research how smart energy and transit technologies can benefit local people and neighborhoods.

www.connectingbristol.org/projects/replicate

IMPROVING ENERGY EFFICIENCY
Boston

Among its many climate-related initiatives, Boston has pledged to lead by example, conducting energy audits of city-owned buildings and exterior lighting and retrofitting them to reduce emissions. Since its launch in 2017, Renew Boston Trust has installed solar panels, heat pumps, and other energy-efficient equipment in police and fire stations, libraries, schools, and other municipal spaces.

boston.gov/environment-and-energy/ renew-boston-trust

INSPIRING CITIZEN CARETAKERS
Athens

As part of an effort to make Athens cleaner and greener, leaders invited individuals, businesses, foundations, and others to sponsor streets, parks, playgrounds, and other public spaces. Since the program's launch in 2019, private and public partners have funded projects including lighting installation, graffiti removal, and tree planting. The program also supports social services and education.

adoptathens.gr/en

PUTTING CARS IN THEIR PLACE
Oakland

When the COVID pandemic hit in 2020, cities had an unexpected opportunity to rethink how residents connected with public spaces. Oakland was one of the first cities to close streets to cars, creating miles of open space for pedestrians and bicyclists to explore— and inspiring dozens of cities around the world to follow suit. The Slow Streets philosophy is still informing the city's approach to neighborhood streets.

oaklandca.gov/projects/oakland-slow-streets

SETTING NEIGHBORHOOD HOUSING TARGETS
Washington, DC

With residents spending more income on housing than ever, Washington, DC became the first US city to set affordable housing targets by neighborhood. The step was part of a housing equity plan intended to ensure distribution of affordable options throughout the city and combat the tendency to cluster affordable housing in low-income areas.

housing.dc.gov

ATTRACTING IMPACT INVESTORS
Phoenix

The desert city developed one of the nation's first Green and Sustainability Bond Frameworks to attract new impact investors interested in supporting environmentally sound projects. In 2020, the city's first Sustainability Bonds issuance—$127 million for water infrastructure—was four times oversubscribed.

www.phoenix.gov/sustainability

ZONING FOR AFFORDABILITY
Cambridge

Confronting soaring housing prices in 2020, the Cambridge City Council took a step that gained national attention: passing a citywide 100 percent affordable housing overlay. This ordinance allows more density than base zoning does—*if* the proposed project is 100 percent affordable—and streamlines the approval process.

cambridgema.gov/CDD/housing/ housingdevelopment/aho

REEMBRACING NATURAL ASSETS
Cleveland

Like many industrial cities, Cleveland relegated its waterfront to highways, train tracks, and factories. Now the city is reembracing its shorelines, exploring projects including a land bridge connecting the downtown to Lake Erie and a riverfront park along the Cuyahoga. With a boost from federal rescue funds, the city is aiming to revitalize disinvested neighborhoods, create new destinations for visitors, and provide wildlife habitat.

clevelandnorthcoast.com

CAPTURING LAND VALUE
Bogotá

The city included land value capture as an essential tool in its revised master plan. In this financing approach, used for more than a century in Colombia and gaining momentum in cities around the world, developers contribute to the costs of infrastructure and other municipal improvements that make their projects possible and profitable. The goal: public benefit from private gains.

lincolninst.edu/publications/multimedia/land-value-capture-explained

SUPPORTING SINGLE MOTHERS
Birmingham

With a $500,000 grant from the national Mayors for a Guaranteed Income network, Birmingham piloted a program that provided 110 households headed by single mothers with $375 per month for a year. Mayor Randall Woodfin characterized the program, which attracted thousands of applicants, as a step toward building a stronger city.

birminghamal.gov/embracemothers

INCENTIVIZING ELECTRIFICATION
Burlington

Intent on reaching net zero by 2030 and doing its part to combat climate change, Burlington provides a wide array of electrification incentives through its city-owned utility, Burlington Electric. Residents can qualify for rebates on purchases ranging from heat pumps to electric snowblowers to car-charging stations.

burlingtonelectric.com/2023rebates

LEVYING A VACANCY TAX
Berkeley

In 2022, Berkeley voters approved a residential vacancy tax intended to combat speculators who were leaving property vacant—and to make housing available for more residents. Measure M levies an annual tax of $3,000 on empty condos, duplexes, and single-family homes not used as primary residences and $6,000 for all other empty units. Along with a similar tax enacted in San Francisco, the tax is one of the first of its kind in the country.

vacancytaxberkeley.org

APPOINTING A CHIEF HEAT OFFICER
Freetown

As part of a global effort by the Arsht-Rockefeller Foundation Resilience Center, Freetown named a chief heat officer tasked with identifying and managing climate adaptation projects. Other participating cities include Athens, Dhaka, Melbourne, Miami, and Santiago. One of the first projects in Freetown was the installation of shades in the city's markets to protect vendors from extreme heat.

onebillionresilient.org/project/chief-heat-officers

ENTERING THE METAVERSE
Seoul

The city of Seoul is making it easier for its tech-savvy population to interact with the government by developing a virtual platform hailed by *Time* magazine as one of the best inventions of 2022. The tool not only provides a way to conduct municipal business—paying taxes, attending meetings, and filing complaints—it also allows users to visit tourist destinations, play games, enter contests, and virtually interact with each other.

metaverseseoul.kr

INVESTING IN AFFORDABILITY
Cincinnati

Besieged by out-of-state investors who purchased thousands of local properties, then proceeded to neglect the houses and the people living in them, the city partnered with the Port of Cincinnati, a regional economic development agency, to buy and renovate more than 190 single-family homes. The hands-on move earned national media attention.

www.cincinnatiport.org

PLANNING IN PHASES
Delhi

On track to become the world's largest metropolitan area, Delhi is suffering the visible ills of rapid growth, from smog-choked streets to looming landfills. These aren't problems one leader or administration can fix. But the city is committed to addressing its challenges piece by piece, creating seasonal action plans that guide significant activities from planting trees to measuring pollution to recycling waste.

ddc.delhi.gov.in/our-work/5

IMAGE CREDITS

Cover
Front: Randall Woodfin by Arron Jackson (left); Hanna Gronkiewicz-Waltz by Rafał Motyl (right); Claudia López by Bogotá Mayor's Office (inside). Back: Muriel Bowser by Khalid Naji-Allah/Washington, DC Mayor's Office (left); Oh Se-hoon by Sipa USA/Alamy Stock Photo (right); Kostas Bakoyannis by Athens Mayor's Office (inside).

Syracuse
10: Syracuse Mayor's Office; 13: Debra Millet via Shutterstock; 14: Detroit Publishing Company Photograph Collection/Library of Congress, Prints and Photographs Division; 15: Philip Scalia/Alamy Stock Photo; 16: Image by Trowbridge Wolf Michaels, a Fisher Associates Landscape Architecture Studio; 17: Courtesy of NUAIR.

San Isidro, Lima
19: Courtesy of Manuel Velarde; 20, 22: Dante Piaggio D/ El Comercio (GDA via AP Images); 23: Municipality of San Isidro; 24: rjankovsky/Alamy Stock Photo.

Warsaw
26: Ewelina Wach; 28: Konoplytska via iStock Editorial/ Getty Images Plus; 30: Tramino via iStock Unreleased/ Getty Images; 31: Bim via iStock/Getty Images Plus; 32: OlyaSolodenko via iStock Editorial/Getty Images Plus; 34: Antagain via iStock/Getty Images Plus.

Santa Monica
36: City of Santa Monica; 38: Natalia Macheda via Shutterstock; 40: Lawrence Anderson/Lorcan O'Herlihy Architects (top); Lorcan O'Herlihy Architects (bottom); 41: Randall R. Howard/Frederick Fisher & Partners; 42–43: Civic Wellbeing Partners; 44: Kiewit Corporation; 45: City of Santa Monica.

Bristol
46: Stephen Fulham/Bristol Mayor's Office; 48: Rob Arnold/Alamy Stock Photo; 49: ZED PODS Limited; 50: Adam Gasson/Alamy Stock Photo; 52: S Swenson/Alamy Stock Photo; 54: Anthony Roberts/Alamy Stock Photo; 55: City of Bristol.

Boston
56: Erin Clark/*The Boston Globe* via Getty Images; 59: Gerrit de Heus/Alamy Stock Photo; 60: Cydney Scott/ Boston University Photography; 61: © Stoss Landscape Urbanism; 62: Elizabeth Felicella/Michael Van Valkenburgh Associates.

Athens
64: Athens Mayor's Office; 65: cenkertekin via iStock Editorial/Getty Images Plus; 69: Athens Mayor's Office; 70: rey perezoso via flickr CC BY-SA 2.0; 73: SHansche via iStock/Getty Images Plus; 72: Andrea Bonetti/ SOOC/AFP via Getty Images.

Oakland
74: Jeff Chiu/AP Photo via Alamy Stock Photo; 77: Sergio Ruiz; 78: Josh Edelson/AFP via Getty Images; 81, 82, 83: Sergio Ruiz.

Washington, DC
84: Khalid Naji-Allah/DC Mayor's Office; 87: Xinhua/ Alamy Stock Photo; 88: Jim Lo Scalzo/EPA-EFE/ Shutterstock; 90: Courtesy of Segregation by Design; 91: Courtesy DLR Group, © Kevin G. Reeves Photographer; 93: Patrick Cavan Brown.

Phoenix
94: Office of Mayor Kate Gallego; 97: 4kodiak via E+/ Getty Images; 98: Reuters/Carlos Barria; 100: Valley Metro; 101: AP Photo/Ross D. Franklin File; 103: DougVonGausig via iStock/Getty Images Plus.

Cambridge
104: Luis Sinco for *Los Angeles Times* via Getty Images; 106: Jorge Salcedo/Alamy Stock Photo; 108: © Damianos Photography (left); © Robert Benson Photography (right); 110: Kyle Klein Photography; 112: Anton Grassl; 113: Photographer: William Horne; Architect: William Rawn Associates in association with Arrowstreet Architecture & Design.

Cleveland

115: portrait artist Robert Hartshorn/image provided by Frank G. Jackson; 117: James E. Walker Jr. via flickr CC BY-NC 2.0; 119: Detroit Publishing Company Photograph Collection/Library of Congress Prints and Photographs Division (top), k_e_lewis via flickr CC BY-NC-ND 2.0 (bottom); 120: Mapping Inequality/University of Richmond; 122: terry vacha/Alamy Stock Photo; 123: drnadig via iStock/Getty Images Plus.

Bogotá

124: Bogotá Mayor's Office; 127: Ron Giling/Alamy Stock Photo; 128: Bogotá Mayor's Office; 131: aRE/Llano Fotografía; 132: Bogotá Mayor's Office (top); mtcurado via iStock/Getty Images Plus (bottom).

Birmingham

134: Daniel Roth/City of Birmingham; 136: Teamjackson via iStock Editorial/Getty Images Plus; 138: City of Birmingham; 139: RGB Ventures/SuperStock/Alamy Stock Photo; 141: Roy Johnson/Alamy Stock Photo (top), LeRoy Woodson/EPA/National Archives (bottom); 142: John Arnold Images Ltd/Alamy Stock Photo; 143: Amanda Bridges-Dunn.

Burlington

144: Glenn Russell/*VT Digger*; 146: Oliver Parini; 148, 150: Burlington Electric Department; 151: halbergman via iStock/Getty Images Plus; 152: Lincoln Brown, courtesy of Duncan Wisniewski Architecture.

Berkeley

155: Office of Mayor Jesse Arreguín; 157: Sundry Photography via iStock/Getty Images Plus; 158: © DLR Group; 160: Marmaduke St. John/Alamy Stock Photo; 161: © Paul Kuroda; 162: Anton Sorokin/Alamy Stock Photo (top), Stars and Stripes/Alamy Stock Photo (bottom).

Freetown

164: Office of Mayor Yvonne Aki-Sawyerr; 166: robertharding/Alamy Stock Photo; 168: Courtesy of Adrienne Arsht-Rockefeller Foundation Resilience Center; 169: Transform Freetown Third-Year Report/City of Freetown; 170: Xinhua/Alamy Stock Photo; 173: Abenaa via iStock/Getty Images Plus.

Seoul

174: Sipa USA/Alamy Stock Photo; 177: Berk Ozdemir via iStock Editorial/Getty Images Plus; 178: Courtesy of West 8; 180: Seoul Metropolitan Government; 181: Fani Kurti via iStock/Getty Images Plus; 182: lukyeee1976 via iStock/Getty Images Plus; 184–185: Gangneung-Wonju National University via ASLA.

Cincinnati

186, 189: Meg Vogel-USA TODAY NETWORK; 190: © Jeffrey Dean; 191: Tomka/Alamy Stock Photo; 193: Cincinnati Enquirer-USA TODAY NETWORK; 195: Cincinnati Parks Foundation.

Delhi

196: Hindustan Times via Getty Images; 199: hadynyah via iStock/Getty Images Plus; 200: paul kennedy/Alamy Stock Photo; 202: NASA Earth Observatory images by Lauren Dauphin, using Landsat data from the US Geological Survey; 205: AP Photo/Manish Swarup.

ABOUT THE AUTHOR

Anthony Flint is a senior fellow at the Lincoln Institute of Land Policy. He is host of the *Land Matters* podcast, contributing editor at *Land Lines* magazine, and a correspondent at Bloomberg *CityLab* and the *Boston Globe*. He is the author of *Modern Man: The Life of Le Corbusier, Architect of Tomorrow* (New Harvest); *Wrestling with Moses: How Jane Jacobs Took on New York's Master Builder and Transformed the American City* (Random House); and *This Land: The Battle over Sprawl and the Future of America* (Johns Hopkins University Press).

ABOUT THE LINCOLN INSTITUTE OF LAND POLICY

The Lincoln Institute of Land Policy seeks to improve quality of life through the effective use, taxation, and stewardship of land. A nonprofit, private operating foundation whose origins date to 1946, the Lincoln Institute researches and recommends creative approaches to land as a solution to economic, social, and environmental challenges. Through education, training, publications, and events, we integrate theory and practice to inform public policy decisions worldwide. We organize our work around three pillars of land policy: fiscal systems; land markets and opportunities; and climate, land, and water. We envision a world where cities and regions prosper and benefit from coordinated land use planning and public finance; where communities thrive from efficient and equitable allocation of limited land resources; and where stewardship of land and water resources ensures a livable future. We work globally, with locations in Cambridge, Massachusetts; Washington, DC; Phoenix, Arizona; and Beijing, China.

LINCOLN INSTITUTE
OF LAND POLICY

ISBN: 978-1-55844-448-5 (Print)
ISBN: 978-1-55844-449-2 (E-Pub)
Library of Congress Control Number: 2023946354

Published by the Lincoln Institute of Land Policy
Chief Content Officer: Maureen Clarke

Editors: Jennifer Sigler and Katharine Wroth
Design: Elizabeth Leeper, Studio Rainwater
Research and Editorial Support: Aliyah Baruchin, Kevin Clarke,
Amy Finch, Hannah Frith, Jon Gorey, Emily McKeigue

Every reasonable effort has been made to identify and acknowledge
the copyright holders of all images reproduced in this book.

The interviews in this book were conducted between 2018 and 2023,
and originally published in *Land Lines*, the magazine of the Lincoln
Institute of Land Policy.

Printed and bound in the US by Puritan Press on FSC-certified stock.

Distributed by Columbia University Press and Ingram Publisher Services.

Lincoln Institute of Land Policy
113 Brattle St.
Cambridge, MA 02138-3400
USA

www.lincolninst.edu